Pulsed Field Gel Electrophoresis

The Practical Approach Series

SERIES EDITORS

D. RICKWOOD
Department of Biology, University of Essex
Wivenhoe Park, Colchester, Essex CO4 3SQ, UK

B. D. HAMES
Department of Biochemistry and Molecular Biology
University of Leeds, Leeds LS2 9JT, UK

★ **indicates new and forthcoming titles**

Affinity Chromatography
Anaerobic Microbiology
Animal Cell Culture
  (2nd Edition)
Animal Virus Pathogenesis
Antibodies I and II
★ Basic Cell Culture
Behavioural Neuroscience
Biochemical Toxicology
★ Bioenergetics
Biological Data Analysis
Biological Membranes
Biomechanics — Materials
Biomechanics — Structures
  and Systems
Biosensors
★ Carbohydrate Analysis
  (2nd Edition)
Cell–Cell Interactions
Cell Culture Models of Epithelia
The Cell Cycle
Cell Growth and Apoptosis
Cellular Calcium

Cellular Interactions in
  Development
Cellular Neurobiology
Clinical Immunology
Crystallization of Nucleic Acids
  and Proteins
★ Cytokines (2nd Edition)
The Cytoskeleton
Diagnostic Molecular Pathology
  I and II
Directed Mutagenesis
★ DNA Cloning 1: Core Techniques
  (2nd Edition)
★ DNA Cloning 2: Expression
  Systems (2nd Edition)
★ DNA Cloning 3: Complex
  Genomes (2nd Edition)
★ DNA Cloning 4: Mammalian
  Systems (2nd Edition)
Electron Microscopy in Biology
Electron Microscopy in
  Molecular Biology
Electrophysiology
Enzyme Assays

Pulsed Field Gel Electrophoresis

A Practical Approach

Edited by

A. P. MONACO

Imperial Cancer Research Fund Laboratories,
Institute of Molecular Medicine,
John Radcliffe Hospital, Oxford

OXFORD UNIVERSITY PRESS
Oxford New York Tokyo

Oxford University Press, Walton Street, Oxford OX2 6DP

Oxford   New York
Athens   Auckland   Bangkok   Bombay
Calcutta   Cape Town   Dar es Salaam   Delhi
Florence   Hong Kong   Istanbul   Karachi
Kuala Lumpur   Madras   Madrid   Melbourne
Mexico City   Nairobi   Paris   Singapore
Taipei   Tokyo   Toronto
and associated companies in
Berlin   Ibadan

Oxford is a trade mark of Oxford University Press

Published in the United States
by Oxford University Press Inc., New York

A catalogue record for this book is available from the British Library

Library of Congress Cataloging-in-Publication Data
Pulsed field gel electrophoresis : a practical approach / A. P. Monaco.
(Practical approach series; 158)
Includes bibliographical references and index.
1. Pulsed-field gel electrophoresis—Laboratory manuals.
2. DNA—Analysis—Laboratory manuals. I. Monaco, Anthony P. II. Series.
QP519.9.P84P854   1995   574.87′3282—dc20   95–4091
ISBN 0 19 963536 6 (Hbk)
ISBN 0 19 963535 8 (Pbk)

Typeset by Footnote Graphics, Warminster, Wilts
Printed in Great Britain by
Information Press Ltd, Eynsham, Oxon

Preface

Pulsed field gel electrophoresis (PFGE) has provided a reliable system for separation of DNA fragments greater than 50 kb. Since its introduction by Schwartz and Cantor in 1984, PFGE has continued to make a major impact on the analysis of both prokaryotic and eukaryotic genomes. PFGE was essential in closing the gap between cytogenetic analysis of mammalian chromosomes which could distinguish deletions in the range of 5 Mb to cloning vectors in *E. coli* which could handle inserts up to 45 kb. PFGE was used to construct long-range (1–10 Mb) restriction maps of mammalian chromosomal regions and was also important in the analysis of both bacterial and protozoan parasite chromosomes. With the introduction of cloning in yeast artificial chromosomes (YACs), PFGE led the way in the construction of YAC libraries, the analysis of individual YAC clones, and the purification of YAC DNA for subcloning in *E. coli*-based vectors or transfer intact to mammalian cells for functional analysis. The theory behind PFGE is discussed in Chapter 1 and many of the practical applications of PFGE are covered in this book with detailed protocols.

Oxford   A. P. M.
August 1994

Contents

Contents

3. Mutation detection and diagnosis using PFGE

*J. T. Den Dunnen, P. Liang, G. J. B. Van Ommen, and
C. Van Broeckhoven*

6. PFGE analysis of yeast artificial chromosomes 119

J. Ragoussis

7. Functional analysis of mammalian genomes using yeast artificial chromosomes 139

Z. Larin

8. PFGE in the study of a bacterial pathogen (*Haemophilus influenzae*)   159

D. W. Hood

9. Analysis of the genomes of protozoan parasites using PFGE   177

David J. Kemp and Roberto Cappai

Contents

Contributors

RAKESH ANAND
Zeneca Pharmaceuticals, Mereside, Alderley Park, Macclesfield, Cheshire
SK10 4TG, UK.

ROBERTO CAPPAI
Walter and Eliza Hall Institute, Post Office, Royal Melbourne Hospital,
Victoria, 3050, Australia.

D. COHEN
Foundation J. Dausset, CEPH, 27 rue Juliette Dodu, 75010 Paris, France.

J. T. DEN DUNNEN
Department of Human Genetics, Leiden University, Medical Genetics
Center Zuid-West Nederland, Wassenaarseweg 72, 2333 AL Leiden, The
Netherlands.

J. K. ELDER
Department of Biochemistry, University of Oxford, South Parks Road,
Oxford OX1 3QU, UK.

M. F. HO
Imperial Cancer Research Fund Laboratories, Institute of Molecular Medicine,
John Radcliffe Hospital, Oxford OX3 9DU, UK.

D. W. HOOD
Molecular Infectious Diseases Laboratory, Paediatrics Department, Institute
of Molecular Medicine, John Radcliffe Hospital, Oxford OX3 9DU, UK.

DAVID J. KEMP
Menzies School of Health Research, PO Box 41096, Casuarina, Darwin,
Northern Territory 0811, Australia.

Z. LARIN
Department of Biochemistry, University of Oxford, South Parks Road,
Oxford OX1 3QU, UK.

P. LIANG
Neurogenetics Laboratory, Born-Bunge Foundation, Department of Bio-
chemistry, University of Antwerp, Universiteitsplein 1, B-2610 Antwerpen,
Belgium.

A. P. MONACO
Imperial Cancer Research Fund Laboratories, Institute of Molecular Medicine,
John Radcliffe Hospital, Oxford OX3 9DU, UK.

P. OUGEN
INSERM U358, Laboratory of Dr M. Lathrop, 12 rue de la grange aux belles, 75010 Paris.

J. RAGOUSSIS
Paediatric Research Unit, UMDS Guy's Campus, London Bridge, London SE1 9RT, UK.

E. M. SOUTHERN
Department of Biochemistry, University of Oxford, South Parks Road, Oxford OX1 3QU, UK.

C. VAN BROECKHOVEN
Neurogenetics Laboratory, Born-Bunge Foundation, Department of Biochemistry, University of Antwerp, Universiteitsplein 1, B-2610 Antwerpen, Belgium.

G. J. B. VAN OMMEN
Department of Human Genetics, Leiden University, Medical Genetics Center Zuid-West Nederland, Wassenaarseweg 72, 2333 AL Leiden, The Netherlands.

PAUL A. WHITTAKER
University Department of Clinical Biochemistry, Level D, South Block, Southampton General Hospital, Tremona Road, Southampton S016 6YD, UK.

Abbreviations

AGE	agarose gel electrophoresis
ARS	autonomously replicating segment
ATCC	American type culture collection
BAP	bacterial alkaline phosphatase
BHI	brain, heart infusion medium
BMD	Becker muscular dystrophy
BSA	bovine serum albumin
cDNA	complementary DNA
CDTA	1,2-cyclohexylenedinitrilotetraacetic acid
CEN	centromere
CEPH	Centre d'Etude du Polymorphism Humain
CFGE	crossed field gel electrophoresis
CGD	chronic granulomatous disease
CHEF	contour-clamped homogeneous electric field
CHO	Chinese hamster ovary
CIAP	calf intestinal alkaline phosphatase
CMT1a	Charcot–Marie–Tooth disease type 1a
c.p.m.	counts per minute
DEAE	diethylaminoethyl
DMD	Duchenne muscular dystrophy
DTT	dithiothreitol
EDTA	ethylenediaminetetraacetic acid
EGTA	ethyleneglycobis (β-aminoethyl)ether tetraacetic acid
FIGE	field inversion gel electrophoresis
FISH	fluorescence *in situ* hybridization
FOLR	folate receptor genes
FSHD	fascioscapulohumeral dystrophy
Hepes	N-2-hydroxyethylpiperazine-N'-2-ethanesulphoric acid
ICRF	Imperial Cancer Research Fund
LINE	long interspersed nucleotide element
LMT	low melting temperature
Mb	mega base pairs
PBS	phosphate-buffered saline
PCR	polymerase chain reaction
PCV	packed cell volume
PEG	polyethylene glycol
PFGE	pulsed field gel electrophoresis
PMSF	phenylmethylsulfonylfluoride
RARE	RecA-assisted restriction endonuclease

RFGE	rotating field gel electrophoresis
RFLP	restriction fragment length polymorphism
rRNA	ribosomal RNA
SDS	sodium dodecyl sulphate
SINE	short interspersed nucleotide element
SLS	sodium lauryl sulphate
SSC	standard saline citrate
STS	sequence tagged site
TAE	tris–acetate–EDTA electrophoresis buffer
TBE	tris–borate–EDTA electrophoresis buffer
TEL	telomere
VSG	variable surface glycoprotein
XK	McLeod's syndrome gene locus
YAC	yeast artificial chromosome

Theories of gel electrophoresis of high molecular weight DNA

E. M. SOUTHERN and J. K. ELDER

1. Introduction

This book is about the practicalities of DNA analysis by pulsed field gel electrophoresis but there are a number of reasons to be concerned with theoretical aspects also. First, the main purpose of the method is to measure the size of large DNA molecules; there is no other convenient way of doing this, and to obtain accurate transformation of mobility to size it is important to understand the relationship between the two quantities. Second, understanding the physical basis of the process may lead to technical improvements; for example, more rapid separations, greater resolution, or separation of bigger sizes. Much has been written about the processes underlying the separation, and it is not the purpose of this chapter to review all of these treatments. Rather, it provides a relatively simple treatment designed to give the reader and practitioner some insight into the workings of the system.

1.1 Origins of pulsed field electrophoresis

Pulsed field gel electrophoresis has remarkable origins. The need to analyse very large DNA molecules arose from a desire to map long genomes. Measurement of DNA molecules by electron microscopy was slow, tedious, and not very accurate, and conventional gel electrophoresis could not be used to analyse molecules longer than about 50 kb.

Zimm (1) had developed a method using viscoelastic recoil—the propensity of long flexible molecules to re-fold to a random coil after they have been stretched by the shearing force exerted on a solution held in the gap between two cylinders, one rotating inside the other. This method, though simpler than electron microscopy, was limited, as it measured only the size of the longest DNA molecule in the population, and it did not provide a way of separating molecules. Schwartz sought to adapt the principle of viscoelastic recoil to gel electrophoresis and with Cantor (2) developed an ingenious electrophoretic analogue of the concentric rotating cylinders. Their idea was based on the reasonable assumption that as DNA molecules are long and thin they must be stretched out when they travel through the small pores of a gel under the influence of an electric field in much the same way as they are stretched by shearing forces. It was further argued that when the field was turned through an angle of 90°, the molecules could not move in the new field direction until they had recoiled to a compact shape and that the time taken for recoil would be greater for longer molecules than for shorter ones. As an inducement to encourage the molecules to find a new path through the gel it was believed that at least one of the two fields should have a gradient of potential (hence the original name for the technique was pulsed field gradient electrophoresis). An experimental system was devised which quickly showed separation of molecules an order of magnitude longer than had ever been separated by conventional electrophoresis, opening up entirely new applications for the technique.

Remarkably, it is now clear that the elegant theory on which the first successful method was based is wrong in one important detail: the mechanism of separation does not depend on time-dependent reorientation of molecules, but rather on forms of movement which produce different rates of progress for molecules of different length. This realization has led to new designs of equipment which give improved performance.

Field inversion gel electrophoresis was discovered by Carle *et al.* (3) during studies of the effect of varying the field angle in pulsed field electrophoresis: it was found that good separation of high molecular weight DNA could be achieved with field angles of 180°, provided the forward and reverse pulses were unequal in strength or duration. A number of theories have been put forward to explain the basis of the effect, and one is included in this chapter.

2. Factors affecting the behaviour of DNA in gels

Two major factors determine the way in which DNA behaves as it travels through a gel, propelled by the electrical forces on its charged backbone: its own structure, and the structure of the gel. This chapter stresses the importance of understanding the mechanism by which the chain penetrates the gel as a key to understanding the separation process.

2.1 Structure of agarose gels

Agarose is a soluble fraction of agar, comprising an alternating copolymer of 3-linked β-D-galactopyranose and 4-linked 3,6-anhydro-α-L-galactopyranose residues (4). X-ray diffraction studies (5) suggest that the polymer has a double helical structure, presumably held together by hydrogen bonds. The stiffness of the gel is accounted for by the formation of bundles of fibres, each comprising some hundreds of double helices (5–7). Arnott *et al.* (5) suggest that 'an accumulation of agarose into a separate "network phase" in a gel which may contain up to 100 parts of water for each part of agarose would leave relatively large voids through which large molecules and particles could diffuse.' A structure comprising thick bundles of fibres with large voids is confirmed by examining gels in the electron microscope (6–8, unpublished observations of EMS) and continuity between voids is indicated by the observation that spherical particles up to 200 nm in radius can be forced through agarose gels by electrophoresis (9). These studies suggest that the radii of pores range from 36 nm for 4% agarose, through 100 nm at 1.5%, to 230 nm at 0.45%; it should be remembered that the fibre bundles comprise many, perhaps hundreds of agarose chains, so that this phase of the gel has a thickness considerably greater than that of the DNA double helix. Electron micrographs indicate that the bundles may have holes in them that are much smaller than the larger voids.

These structures and dimensions become important for the discussion of ideas of how DNA molecules progress through the gel under the influence of an electric field.

2.2 DNA in solution and in gels

The DNA double helix is flexible. Different sequences have a different degree of flexibility, and certain sequences induce bending in the chain. These features have effects on gel mobility. However, in long chains, which are the prime concern of this book, it is likely that they will be averaged out and swamped by the effects of chain length. As DNA appears frequently to travel through gels led by a loop, it is important to note that there is a limit to the curvature to which the DNA double helix can be bent, corresponding to about 150 bp on average. This corresponds to an arc with diameter of *c*. 30 nm, approximately the dimensions of a pore in concentrated agarose gels.

Yanagida (10) observed the shape and movement of long DNA molecules stained with a fluorescent dye by video microscopy. They showed that molecules alternate rapidly between conformations which they describe as flexible rods, spheres, and ellipsoids. Smith *et al.* (11) applied the same method to DNA in gels. In the absence of an electric field, DNA of phage λ appeared as stationary blobs *c*. 2 μm across—close to the expected random coil diameter of 1 μm. Under the influence of an electric field, 'Molecules in the gel underwent a constrained motion, winding through invisible pores in the gel, elongating in the direction of the field, and then contracting as the tails caught up with the heads.' Other figures seen in DNA as it moved through the gel were inverted U shapes, suggesting that molecules became wrapped round an obstacle with the two ends pulled forward so that the molecule extended to its full contour length. When this happened, the molecule 'slipped free of the obstruction and moved into the path of the longer arm.' Longer DNA, from yeast chromosomes, showed more extended configurations aligned with the electric field. After removing the field, molecules shrank back to fill the pores, showing in a clear way both the structure of the gel and the elastic properties of long DNA molecules.

Thus the structure of gels and the behaviour of DNA in solution are well enough understood at a gross level to incorporate into models that predict the rules governing the relationship between mobility and size.

3. Relationship between DNA size and conformation, and mobility: experimental observations

3.1 Conventional constant field gel electrophoresis

The mobility of linear DNA in agarose gels under constant field is inversely proportional to its size and a simple formula, based on the reciprocal relationship, can be used to calculate DNA size from mobility (12, 13). Above a certain size limit which depends on the gel concentration and voltage gradient, all molecules have the same mobility (henceforward referred to as the limit mobility). The techniques of pulsed field gel electrophoresis were developed to resolve the fragments above this size limit.

Also relevant to understanding the mechanism of gel electrophoresis are the findings that in relatively concentrated agarose the mobilities of long supercoiled or linear molecules may exceed those of shorter ones (ref. 14 and unpublished experiments of C. Tyler-Smith); that supercoiled circles run faster than linear molecules of the same length, but open circles run slower; and that open circles may come to a stop at high voltage gradients (15).

3.2 Field inversion gel electrophoresis: FIGE

The attraction of FIGE as a means of separating long DNA molecules is that it requires very simple apparatus which, unlike that used for crossed field electrophoresis (Section 3.3), can be readily constructed from conventional equipment (3). In its simplest form, it requires only a switching device to reverse the polarity of the electrodes periodically. Field inversion protocols vary: the forward pulse may be longer in duration or of higher voltage than the reverse; different pulse times are used according to the size range to be separated; the ratio of forward to reverse pulse times may be different in different protocols (16); and ramping, a gradual increase in the pulse times, is often used to extend the range of separation.

FIGE has little effect on the separation of molecules which would be resolved by conventional constant field electrophoresis. The major effect is on those molecules of high molecular weight which would travel with the limit mobility. The relationship between size and mobility during FIGE is complex. There is a range in which mobility decreases with size, but there is an inflection point above which longer molecules travel faster than smaller ones (16). This complex relationship makes it difficult to obtain accurate sizes from FIGE, as two molecules with quite different sizes can have the same mobility; it is for this reason that most users prefer pulsed or crossed field electrophoresis.

4

3.3 Pulsed and crossed field gel electrophoresis: PFGE and CFGE

Several systems have been developed for separating very large DNA molecules—up to *c.* 7 Mb. The shared characteristic is a means of driving the DNA molecules in directions at an obtuse angle to each other in alternating pulses. Most devices achieve this by switching between two or more sets of electrodes. An alternative method uses one set of electrodes and moves the gel through a preset angle between pulses. Early theory of how the separation worked required inhomogeneous fields (2). However, these systems produced gel tracks with a high degree of distortion which made it difficult to compare mobilities between tracks, precluding accurate size measurements. More recent systems use homogeneous fields which produce straight tracks (17, 18). Again the relationship between size and mobility is complex, though without the gross instabilities found in FIGE. As with FIGE, PFGE and CFGE have little effect on the mobility of fragments below the limit mobility of conventional gel electrophoresis; the effect of the procedure is to slow down those molecules which move at the limit mobility and to move the limit mobility to a larger size. Within a size range dependent on the pulse regime and voltage, the separation between molecules is roughly proportional to the size difference; between this size range and the new limit mobility is a range in which the separation is much greater. This latter effect has two practical consequences: it makes size measurements inaccurate in this range unless there is a sufficient density of size markers, and it gives good resolution of fragments even if molecules are close in size. Thus for optimal separation or accurate sizing it is important to know the characteristics of the system used for separation.

4. Models, theories, and simulations

4.1 Constant field electrophoresis

It is important to understand how DNA moves through a gel under conventional electrophoresis: this process is not only important in its own right, but each step in discontinuous methods such as FIGE and PFGE must involve processes similar to those which operate in continuous electrophoresis. There have been several attempts to find a theory of continuous gel electrophoresis. They can be classified into three main groups.

(a) The DNA molecule is treated as a random coil which behaves as a globular molecule interacting with the gel in accord with Ogston's model (19). This model is unsatisfactory even for small DNA molecules as it produces a semi-logarithmic relationship between size and mobility rather than the reciprocal relationship that is observed.

(b) The biased reptation model (20, 21) represents the system as a 'snake moving through grass'. In this theory, the DNA molecule is thought of as occupying a notional convoluted tube in the gel, through which it progresses led by one end; the forces acting to retard the progress of the molecule are the frictional interactions with the sides of the tube. Although this model successfully explains the reciprocal relationship between mobility and size, it does not explain some features of the system such as the fact that molecules above a certain size travel at the same rate or may even travel more rapidly than smaller molecules under some conditions.

(c) A third group of theories treats the DNA molecule as a long, highly flexible chain which travels through the gel in fits and starts, becoming entangled, trapped, and hooked on the fibres, piling into tangled bundles in the vacuoles. This is the behaviour observed by video microscopy. One of the more persuasive models is that of Deutsch (22) who suggests that DNA molecules travelling through the gel adopt two main forms: the molecules move in a stretched conformation, but occasionally, the front end becomes compressed, and then trapped in a pore. When this happens, the rest of the molecule piles up behind it, and the molecule is stuck in the pore until loops or ends find a new path out of the tangled molecule in the pore and initiate a new movement in the stretched state. We will show how a treatment of conventional gel electrophoresis based on a simple version of this model fits well the main features of the system. An adaptation to FIGE also fits well the peculiar features of this method. An even simpler model can explain the main features of the 'crossed field' methods.

There are two ways in which we can use models to predict the effects on mobility of various parameters, such as DNA length or gel concentration. One approach is to derive mathematical expressions relating mobility to the relevant parameters and the other is to use computer simulations. Both of these methods have been used with some success and we will show examples of their application to conventional gel electrophoresis, to field inversion, and to pulsed field separations.

For computer simulations and to develop analytical equations, Deutsch (22) treated the DNA molecule as a chain made up of beads on freely hinged links and successfully reproduced some of the main features of the process. In unpublished work, we used a simpler model with some features in common with that of Deutsch, but which treats the DNA molecule as a flexible chain with charge uniformly distributed along it. The most important feature of this model is the way in which we envision the structure of the gel; other models consider the gel as comprising voids, pores, or tubes within a mesh of fibres which are impenetrable to DNA molecules. In our model, the gel is treated as a three dimensional reticulum, with large pores or vacuoles, within bundles of fibres, and we propose that the bundles have small holes which are penetrable by the ends of linear DNA molecules but not by loops or circular molecules, which can only move through the larger pores. There is no direct evidence for this proposal, but everyday experience, for example threading a needle, suggests this behaviour. A model based on these proposals is easy to formulate and to simulate in the computer. We have explored its properties and found that it reproduces many of the properties of both conventional gel electrophoresis and field inversion gel electrophoresis.

4.1.1 Assumptions of the model

Consider first the movement of DNA molecules travelling through gels under a constant uniform field, as in conventional gel electrophoresis, and make the following assumptions.

1. Linear DNA molecules with a contour length greater than the average pore size of the gel can only make forward progress through a gel in one of two extended states:

   (a) *led by one end,* with the other pointing backwards;

   (b) *led by a loop* pointing in the direction of motion, with both ends trailing (*Figure 1*).

   All other extended states include a *hook* (concave loop) which will inevitably become caught on a gel fibre, and must be resolved into one of the forms already described before the molecule can make progress (this can be seen in *Figure 2* of ref. 11).

2. The pores in the gel are highly non-uniform. The structure seen in the electron microscope resembles a bath-tub sponge: a network of dense bundles of fibres, with large voids. Assume that some pores, those within the bundles for example, are small enough to allow ends, but not loops, to penetrate.

3. When an end penetrates such a pore, its rate of movement is slowed significantly (*Figure 1a, i*). When an end becomes trapped in this way, the trailing strand pushes past the trapped front end forming a loop in an adjacent large pore (*Figure 1a, ii*). The compact states predicted by Deutsch (22) and seen by Smith *et al.* (11) would have to be resolved into an extended state in order to penetrate the gel. Such states may be transients formed when the chain piles up in a pore after the front end is trapped in a fibre bundle, and in our model, this state would simply be an intermediate between the trapping of the end and the start of the formation of a loop round the fibre in which the end is trapped. Thus the formation of piles has no effect on mobility in our treatment.

4. The loop continues to progress by finding a path through the large pores. During this process the trapped end may remain trapped and so the loop, which forms at this end, progresses along the chain towards its free end, as it moves forward in the gel (*Figure 1a, ii* and *iii*). The free arm of the

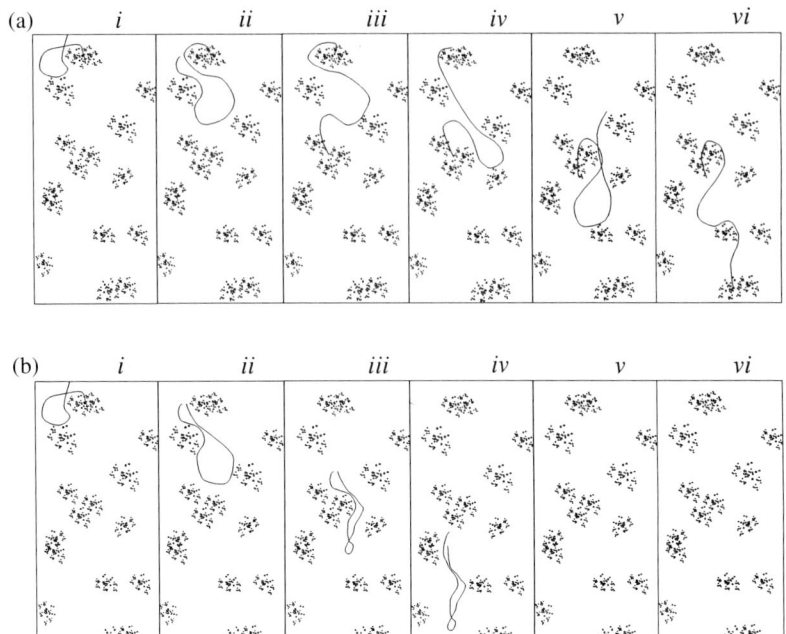

Figure 1. Movement of a DNA chain through an agarose gel under the influence of a continuous field. (a) (*i*) a molecule with one end penetrating a fibre bundle; (*ii*) a loop forms by the trailing strand pushing past the trapped end; (*iii*) the molecule inverts so that the back end is now the leading end; (*iv*) the new leading end becomes trapped and the other end disengages from the fibre bundle; (*v*) and (*vi*) the process is repeated. (b) In this case, which represents the behaviour of long molecules or of shorter molecules in high fields, both ends become untrapped and the molecule is led by a loop which can not penetrate a fibre bundle.

loop may eventually overtake the end tethered in the fibre. A short molecule will extend until the loop is resolved and the molecule is fully extended but has turned around, in a 'whiplash' movement.

5. For simplicity in the computer simulation assume that, once fully extended, the trapped end detaches from the small pore and moves freely through the gel again at the untrapped rate. But the molecule may again become trapped (*Figure 1a, iii* and *vi*).

6. Make the assumption that long molecules may detach from the small pores once the loop has reached a length, the critical length, at which the force on the trapped end is sufficient to tug it free. Once a looped molecule has detached from a trap, it travels forward in the gel and cannot become trapped again (*Figure 1b*). (It is this feature of the model that makes long molecules all travel at the same rate and possibly faster than shorter ones in the simulations of conventional gel electrophoresis.)

8

4.1.2 Simulations of conventional gel electrophoresis

The movement of molecules in a gel in the way described above was simulated in the computer. The main parameters affecting the behaviour of the DNA molecules were the following.

1. The *density of traps*, which was given a Poisson distribution.

2. The *relative rates of end and loop migration*. These were varied; it was assumed that in the untrapped state, molecules led by an end travel at the same rate as, or faster than, those led by a loop. The forward rate of a loop with one end trapped was assumed equal to that of a loop with both ends free. However, as the trapped end holds one half of the chain stationary, the *net* rate of movement of the molecule, which is the rate at which the loop progresses along the chain, is half that of a free looped molecule.

3. The *critical length* at which a loop would pull an end from a trap.

Other parameters to the program were the number of molecules in each size class, the size of the monomer of the series, the number of highest *n*-mer in the series, and the duration of the run.

4.1.3 Comparison of simulations with results

The main features which characterize conventional gel electrophoresis are seen in the simulations (*Figure 2*) described below.

1. *The inverse relationship between DNA length and mobility.* Within a certain range of sizes the mobility of a DNA molecule during gel electrophoresis is inversely proportional to its length (12). This relationship emerges in a natural way from the model: a molecule travels rapidly through the gel at a rate which is independent of length until its end becomes trapped. Once trapped, the time taken for the molecule to untrap is determined by its length; it is the time taken for the loop to progress from the trapped end to the other, which is proportional to the length of the molecule. As with traffic in a crowded city, journey time is determined mainly by delays rather than velocity during the periods when movement is unimpeded.

2. *The limiting mobility above a certain size.* One of the oddest features of the behaviour of DNA during gel electrophoresis is the uniform mobility of fragments above a certain size. Few models explain this feature. It emerges from our model as a consequence of the formation of loops: long molecules are more likely to detach from traps as loops than short ones, which are more likely to whiplash round to a fully extended state before detachment (*Figure 1a* and *b*).

3. *Inversion in the size/mobility relationship at high field strengths.* It has been shown that long molecules can overtake shorter ones at high field strengths (14). This unexpected feature arises from the model because

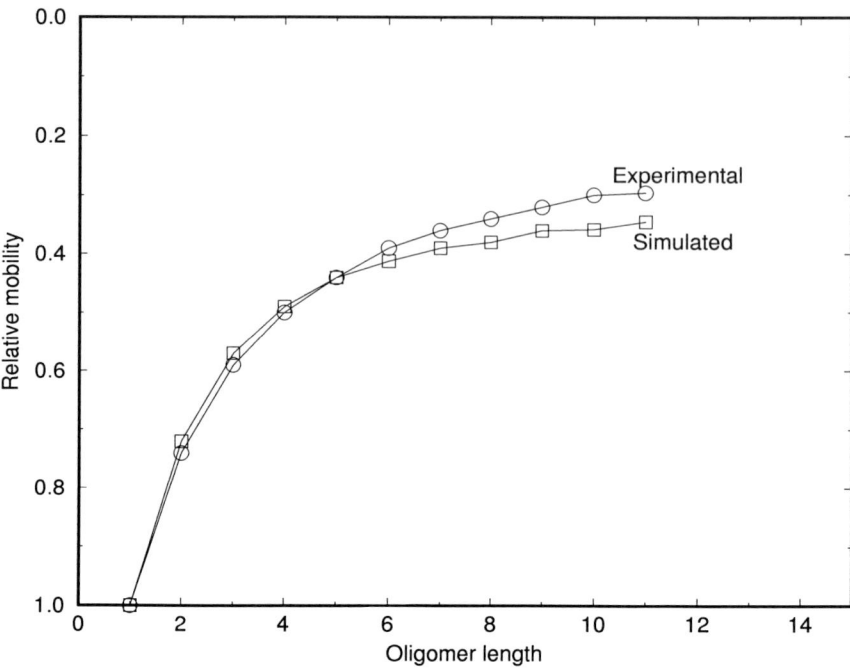

Figure 2. Comparison of experimental data for DNA molecules separated by constant field gel electrophoresis with computer simulated data. The experimental data are taken from ref. 24, in which an oligomeric series made by ligating the plasmid pAT153, monomer length 3657 bp, was separated in a 0.4% agarose gel for 48 h at 0.75 V/cm. This work showed that the upper limit for useful separation of DNA fragments by constant field electrophoresis is around 50 kb. Mobilities were measured to a high accuracy using a digitizing densitometer. Computer simulations carried out as described in the text show remarkably close fit to the experimental data. Note that relative mobilities are accurately predicted, and that the 'limit mobility' of the large molecules which are not separated is also predicted.

The model thus provides a theoretical basis for the observed relationship between size and mobility for continuous field electrophoresis. This has practical implications for the choice of conditions for optimizing separation, but more importantly, for accurate estimation of size from mobility, as is essential for many applications of gel electrophoresis: the reader is referred to ref. 13 for a more extensive discussion of molecular weight determination, and for practical help in choosing and using the appropriate relationship.

molecules which are travelling led by a loop cannot become trapped; longer molecules spend very little time in the trapped state, whereas short molecules become trapped frequently as they 'tumble' through the gel. Since at high field strength the rate of loop penetration will increase, relative to the rate of untrapping, long looped molecules will travel faster than short molecules which spend more time in the trapped state.

4.2 Field inversion gel electrophoresis: FIGE

The most important extension of the model as applied to FIGE is that when the field is reversed, any DNA molecule which is travelling forward as a loop, whether trapped or not, will take off in the reverse direction led by two ends (*Figure 3*). These ends are not bound together and so can take off in separate directions; the DNA between the ends will then quickly become snagged round a fibre, forming a 'hook'. How these hooks are resolved depends on the length of the molecule and determines their mobility, as shown in computer simulations, which were run as follows.

1. Assume that all DNA molecules are fully extended at the point of loading on the gel.

2. On the first pulse, which is a forward pulse, DNA molecules behave as described in the previous section for continuous electrophoresis, and the process of initiating a loop is assumed to be the same for all lengths of molecule.

3. At the end of the forward pulse, however, the configuration of a loop will depend on the length of the molecule. *The proportion of the molecule which is in the two arms of the loop differs for molecules of different length*:

   • a short molecule may have time to whiplash completely round the trap, ending up in a stretched state (*Figure 3a, ii*);

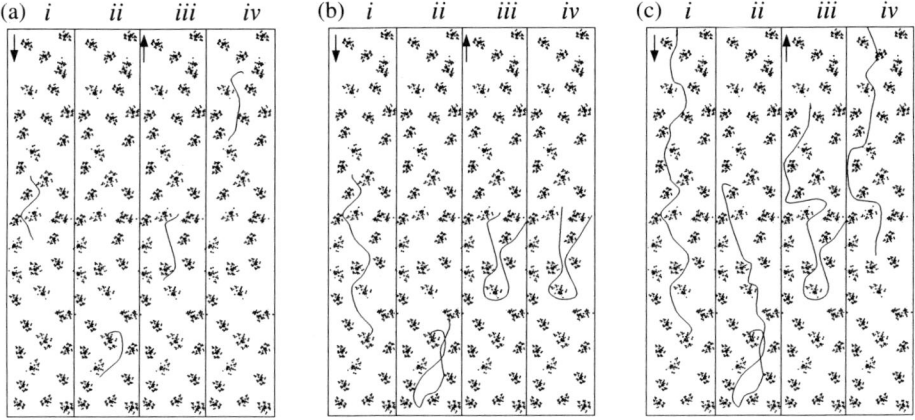

Figure 3. Model for movement of a DNA chain through an agarose gel under the influence of FIGE. Panels (a), (b), and (c) represent the behaviour of a short, medium-sized, and long molecule of DNA moving through a gel according to the model described in detail in the text. The main feature determining the mobility of the molecule is shown in column *iii* of each panel. The short and long molecules are able to move on the reverse pulse, but the medium-sized molecule is immobilized because it hangs over a fibre, and the pull on each arm is equal.

11

- a molecule of intermediate length may end up with arms of approximately equal length (*Figure 3b, ii*);
- a long molecule will have one long and one short arm at the end of the pulse (*Figure 3c, ii*).

4. On reversing the field:

- all trapped ends are released and loops become hooks (*Figure 3b, iii* and *c, iii*);
- hooked molecules hang around fibres at the position in the chain of the loop formed in the previous pulse;
- a hook is resolved by its longer arm tugging the shorter round the snag;
- the pull, and therefore the rate at which the molecule unwraps itself, is proportional to the difference in length between the long arm and the short arm. Therefore a hook with equal arms is stuck, because the force on the two arms is equal (*Figure 3c, iv*);
- once a hook is resolved, the molecule moves in the gel with the rate defined for linear molecules, which is independent of length. The end may now again become trapped and form a loop pointing in the direction of the reverse field.

All subsequent pulses, forward and reverse, are treated in the same way.

4.2.1 Comparison of simulations with experiments

Simulations were run varying the parameters as for continuous electrophoresis, varying also the length of the forward and backward pulses. The program also allowed ramping of the pulse length. Some results are shown in *Figure 4*. The features to note are as follows.

1. Molecules above the size which travel at the same rate in simulations of continuous electrophoresis are separated in the FIGE simulations.

2. The size range of molecules which are separated in the FIGE simulations increases with the pulse time.

3. There is an inflection in the relationship between mobility and size. The inflection point moves to bigger sizes with longer pulses.

Thus many of the complex features of FIGE are reproduced in these simulations.

The value used for the mobility of a molecule between traps was estimated from that of low molecular weight DNA in dilute agarose gels. The relative rates for untrapped loops/linears were varied between 1:1 and 1:5, which covers the experimentally observed range of relative rates of linear to circular forms. Changing these ratios had a marked effect on the results, but the general features were similar in all cases.

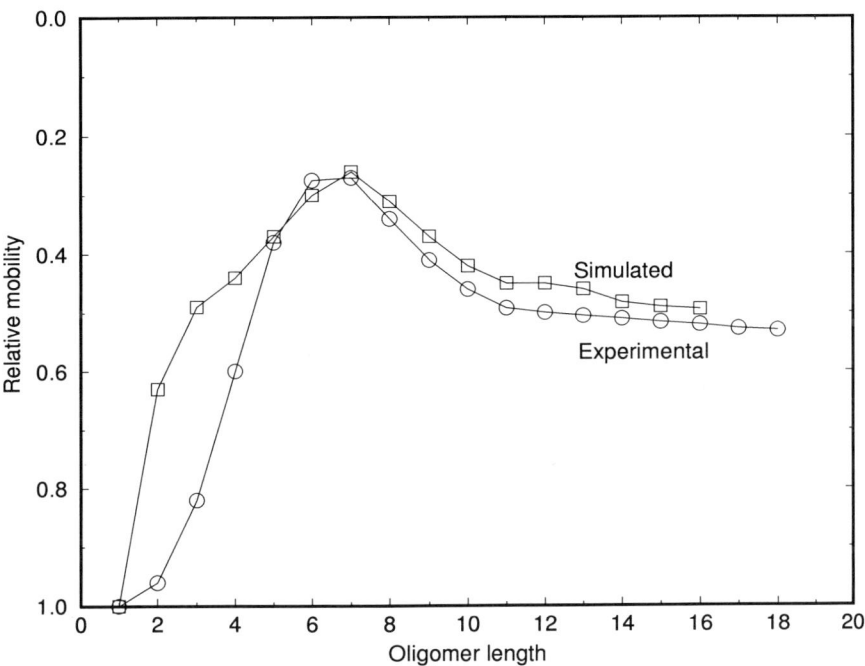

Figure 4. Comparison of experimental data for DNA molecules separated by field inversion gel electrophoresis with computer simulated data. The experimental data are taken from ref. 16. The measurements were performed by two-dimensional separation of an oligomeric series of phage λ DNA, so that the monomer is 48.5 kb and the longer molecules are integer multiples of this size. The first separation was carried out using a crossed field method, and the second using FIGE in which the forward to reverse pulses were in the ratio 3:1. The reader is referred to the original paper for further details, and for a description of a wide range of protocols to give optimum separation by FIGE for particular applications. The simulated data were produced as described in the main body of the text, using the following parameters: forward pulse 4.5 s, reverse 1.5 s; rate of penetration by a molecule led by a free end 50 μm/s, rate of molecule led by a loop 15 μm/s; fibre 'traps' occur at one per 20 μm; monomer persistence length 4.0 μm.

The general shape of the simulated curve is remarkably similar to that of the observed data. Bostock noted three types of behaviour for molecules in the different size ranges: in the low size range, separation is close to being proportional to the difference in size between the fragments; above this range is a short compression in which molecules of different size all have the same mobility; above this range, up to the largest fragments analysed, there is an inversion, in which larger fragments travel faster than smaller ones.

All of these features are reflected in the computer simulations, and the transitions from one behaviour to another occur in the same size range as the observed transitions. Furthermore, in other simulations not shown here, it was found that changing parameters, such as the length and ratios of the forward and reverse pulses, produced trends similar to those found in experiments. This suggests that the broad assumptions of the model are correct, providing a theoretical framework for understanding the underlying mechanism of separation with implications for the design of experiments.

5. Theory of crossed field electrophoresis

Several different designs have been developed for equipment which separates long DNA molecules by driving the molecules alternately in directions at oblique angles (23). We coined the term crossed field electrophoresis to distinguish these methods from FIGE, because we believe the underlying principle of the separation to be somewhat different. Early theories were based on the notion that the separation is based on the DNA molecules reorienting between pulses in a time-dependent process which varied with the length of the molecule. These theories find little support in experiment, but a mechanism which is based on length-dependent molecular ratcheting fits the data quite well. This simple theory starts from the assumption that DNA molecules must adopt a stretched conformation as they pass through gels in which the pore size is much smaller than the contour length of the molecules. At the end of each pulse a molecule will be left stretched in the orientation of the field. It is postulated that if a new field is applied at an angle oblique to the orientation of the previous field, the molecule is more likely to take off by a movement which is led by what was its back end. The cumulative effect of this ratcheting motion is to subtract from the molecule's forward motion at each step, an amount which is proportional to its length. The configurations predicted by this model have been seen by fluorescence microscopy (11), and it fits many of the features of the separation. In its original formulation, it was assumed that molecules were led through the gels by a physical end; it now seems that they may also be led by loops. This makes the ratcheting mechanism even more likely, as the loop at the front end will be snagged when the field is changed, so that movement of the molecule must be led by the trailing, free end. The effect of looping on the quantitative predictions of the model will be small, provided that the average length of a looped molecule is proportional to contour length, as seems likely. On the other hand, looping may well provide a basis for the explanation that all molecules above a certain size do penetrate the gel at a limit mobility. This observation is not explained by the simple theory.

The path of an extended DNA molecule in crossed field gel electrophoresis

As we have seen, all DNA molecules above a certain size pass through a gel by conventional electrophoresis at approximately the same rate. If they are extended their back ends will trail behind their front ends at distances which depend on the lengths of the molecules. If the direction of the field is then changed in such a way as to push the DNA molecule sideways, it can only begin to move by leading off with one of its ends, or by forming a new 'end' by kinking. Movement led from within the molecule requires more work and probably occurs less often than movement led by an end. If the new field direction is at an acute angle to the old, movement will be led by the front end. But if it is greater than 90° it will be led by the old back end for two reasons; first, because leading with the old front end would turn the molecule through a sharper angle, which would require more work; and second, because the old front end is likely to be looped, as explained above. As the back ends of the molecules were left at different positions in the previous pulse, the starting point for the new movement is different for molecules of different length (*Figure 5*).

This model predicts that if the angle between the fields is less than 90°, DNA molecules will behave as they do in a gel that is run conventionally,

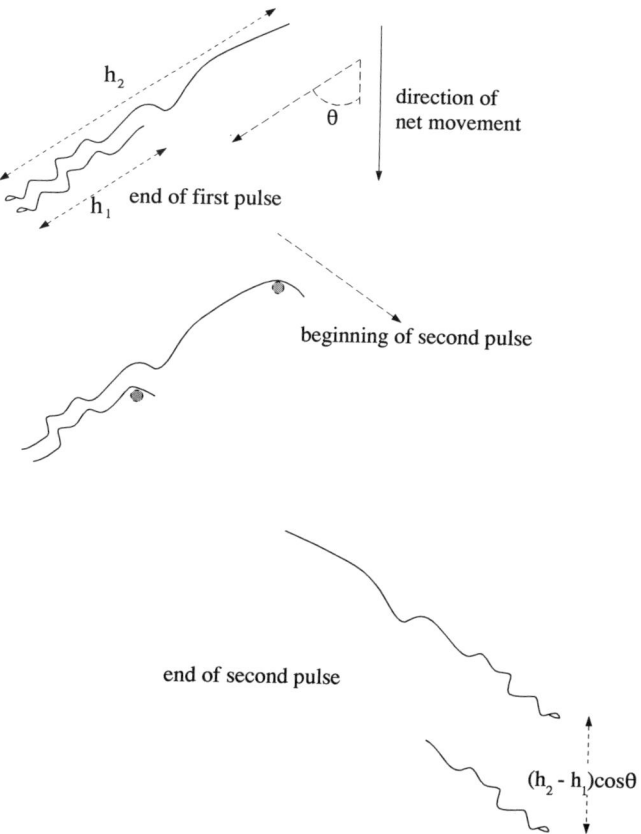

Figure 5. Movement of a DNA chain through an agarose gel under the influence of PFGE. A long and a short molecule are shown at the end of the first pulse, stretched out, with their front ends together. On switching the field through an angle greater than 90°, the back ends initiate movement through the nearest pore. At the end of the second pulse, the longer molecule has been held back more than the small one by a distance that is proportional to the difference in their lengths. See ref. 18 for more detailed treatment and examples of separations.

i.e. with no turning, whereas if the angle is greater than 90°, they will be held back in the gel at each turn by a distance that is proportional to their length; this prediction was confirmed by experiments which showed that the separation between molecules is roughly proportional to the difference in their lengths. A further prediction of the model is that the separation will be proportional to the total number of pulses and, up to a certain limit discussed later, independent of pulse length.

The model may be expressed more precisely as follows. The distance moved in a single pulse by the front end of any molecule along the direction of the field is

$$d = vt,$$

where v is the velocity of DNA in conventional gel electrophoresis under the same conditions and t is the duration of the pulse. The distance moved by the front end along the axis of the gel is

$$d \cos \theta,$$

where θ is the angle of the field to the axis of the gel. After one pulse, the distance along the axis of the gel between the starting position of the front end of the molecule and the final position of the back end of the molecule is

$$m = (d - h) \cos \theta,$$

where h, the molecule's apparent end-to-end length, is less than or equal to the contour length. After one pulse, the distance separating the back ends of two molecules i and j will be

$$m_i - m_j = -(h_i - h_j) \cos \theta.$$

After N pulses, the net distance moved by a molecule i will be

$$M_i = Nm_i = N (d - h_i) \cos \theta = (D - Nh_i) \cos \theta,$$

where D is the distance moved by DNA in conventional electrophoresis under the same conditions and for the same total time. The separation between molecules i and j will be

$$M_i - M_j = N (m_i - m_j) = - N (h_i - h_j) \cos \theta. \qquad [1]$$

Thus, as observed from experiments, separation is proportional to the number of pulses N. Furthermore, for a given run time T, $N = T/t$. Hence, separation is inversely proportional to pulse length, since there will be more short pulses, and each one leads to the same amount of separation.

Size limit for separation
During a pulse, the distance moved by the back end of a molecule is $d - h$. If h is greater than d, the back end will not progress past the position occupied by the front end on the previous pulse. Thus separation will extend up to

molecules whose apparent length is equal to d. Since d is proportional to the pulse length, the size limit for separation increases in proportion to the pulse length, as is found in experiments.

5.1 Other experimental support for the model of crossed field electrophoresis

Many of the predictions of the above model are met by published observations, which have shown that separation is inversely proportional to the duration of the pulse, and that the limit for separation increases with pulse time. In addition it has been shown that separation occurs at obtuse but not at acute field angles, and that there is no requirement for a non-uniform field, which is an important component of the original model.

Effect of reversing the current before changing the field angle
If turning the field through an obtuse angle does have the effect of swapping the leading end from the front to the back of the molecule, the same effect might be achieved by reversing the polarity of the current, in order to reverse the direction of the molecule, and then turning the gel, with the current still switched on, through an acute angle. This regime did indeed give separation of the λ oligomers, lending further support to the model (*Figure 4* of ref. 18).

Relationship between DNA length and mobility
A striking feature of the band pattern formed by the oligomeric series of λ DNA is the uniformity of spacing between members of the series (*Figures 5* and *8* of ref. 18). The model predicts that a molecule will be held back at each step by a distance proportional to a simple function of its length: the uniformity of spacing of the oligomeric series over a considerable range suggests that the relationship between the end-to-end distance of the DNA and its contour length is close to proportionality. The non-uniform spacing above a certain size is a function of pulse length and is explained below.

Effect of pulse length
The model requires that separation is proportional to the number of pulses (Equation 1). That this is so can also be seen from the data in *Figure 5* of ref. 18, which shows that separation is approximately inversely proportional to pulse time. The model also suggests that the size limit for separation will be proportional to pulse length, as this limit is equal to the distance moved by any DNA molecule during the pulse. The size of the largest λ oligomer in the uniformly separated series is proportional to pulse time.

Molecules above a certain size are not separated
To make forward progress, the back end of a molecule must pass the point occupied by its front end at the start of the pulse (*Figure 5*). Molecules above a certain size will be longer than the distance covered during the pulse and

on the turn will 'hang' in the same position. Decreasing the pulse length brings smaller molecules into this class. The simple model, which assumes that DNA is a flexible rod, cannot explain the fact that long molecules do not hang at the point of application, but penetrate the gel to a considerable distance, as they would in constant field electrophoresis. A more complete model of crossed field gel electrophoresis, which takes into account the highly flexible nature of the DNA chain and which allows movement initiated by internal loops, would be needed to incorporate penetration by very long molecules.

Between those molecules that are not separated and those that are separated by a distance proportional to their length, there is a run of oligomers which are spaced wider than the uniform spacing lower in the size range. Spacing is uniform up to a point, above which it increases for a few steps before it is compressed up to the point where no separation is observed. This pattern can be also explained by assuming that the DNA chains are not rigid, and the end-to-end distance varies from one pulse to another: molecules which are close in length to the distance travelled in a pulse will sometimes be too extended to pass the turning point and therefore on some pulses will make no net progress, decreasing their mobility beyond what is predicted by the model. Computer simulations predict the observed patterns quite closely (JKE and EMS, unpublished data). This extension of separation has important practical consequences; it permits the choice of optimum conditions for separation in a particular size range, and it makes accurate sizing difficult unless markers are close in mobility to the unknowns.

References

1. Zimm, B. H. (1985). In *Methods in enzymology* (ed. C. H. W. Hirs and S. N. Timasheff), Vol. 117, pp. 94–7. Academic Press, London.
2. Schwartz, D. C. and Cantor, C. R. (1984). *Cell*, **37**, 67.
3. Carle, G. F., Frank, M., and Olson, M. V. (1986). *Science*, **232**, 65.
4. Araki, C. and Arai, K. (1967). *Bull. Chem. Soc. Japan*, **40**, 93.
5. Arnott, S., Fulmer, A., Scott, W. E., Dea, I. C. M., Moorhouse, R., and Rees, D. A. (1974). *J. Mol. Biol.*, **90**, 269.
6. Amsterdam, A., Er-el, Z., and Shaltiel, S. (1975). *Arch. Biochem. Biophys.*, **171**, 673.
7. Attwood, T. K., Nelmes, B. J., and Sellen, D. B. (1988). *Biopolymers*, **27**, 201.
8. Waki, S., Harvey, J. D., and Bellamy, R. D. (1982). *Biopolymers*, **21**, 1909.
9. Serwer, P. and Hayes, S. J. (1986). *Anal. Biochem.*, **158**, 72.
10. Yanagida, M., Hiraoka, Y., and Katsura, I. (1982). *Cold Spring Harbor Symp. Quant. Biol.*, **47**, 177.
11. Smith, S. B., Aldridge, P. K., and Callis, J. B. (1989). *Science*, **243**, 203.
12. Southern, E. M. (1979). *Anal. Biochem.*, **100**, 319.
13. Elder, J. K. and Southern, E. M. (1983). *Anal. Biochem.*, **128**, 227.

14. Noolandi, J., Rousseau, J., Slater, G. W., Turmel, C., and Lalande, M. (1987). *Phys. Rev. Lett.*, **23**, 2428.
15. Levene, S. D. and Zimm, B. H. (1987). *Proc. Natl. Acad. Sci. USA*, **84**, 4054.
16. Bostock, C. J. (1988). *Nucleic Acids Res.*, **16**, 4239.
17. Vollrath, D. and Davies, R. W. (1987). *Nucleic Acids Res.*, **15**, 7865.
18. Southern, E. M., Anand, R., Brown, W. R. A., and Fletcher, D. S. (1987). *Nucleic Acids Res.*, **15**, 5925.
19. Ogston, A. G. (1958). *Trans. Faraday Soc.*, **54**, 1754.
20. Lerman, L. S. and Frisch, H. L. (1982). *Biopolymers*, **21**, 995.
21. Lumpkin, O. J., Dejardin, P., and Zimm, B. H. (1985). *Biopolymers*, **24**, 1573.
22. Deutsch, J. M. (1988). *Science*, **240**, 922.
23. Anand, R. (1986). *Trends Genet.*, **2**, 278.
24. Elder, J. K., Amos, A., Southern, E. M., and Shippey, G. F. (1983). *Anal. Biochem.*, **128**, 223.

2

PFGE in physical mapping

M. F. HO and A. P. MONACO

1. Introduction

The development of pulsed field gel electrophoresis (PFGE), which allows
the separation of DNA molecules as large as 10 Mb, has enabled large regions
of genomic DNA to be mapped and analysed without cloning the DNA
(1–3). Its wide separation range has bridged the size gap between cytogenetic
methods (>5 Mb) and conventional gel electrophoresis (up to 50 kb). Prior
to this, genetic mapping has been the primary method for mapping genes
responsible for inherited disorders. Genetic maps relate the apparent distance
between genes or other detectable markers to the relative frequency of
genetic events, such as recombination; hence its resolution is dependent on
the number of available meiotic events and recombination frequency. The
latter not only differs between male and female but varies for different
regions of the genome. In contrast, physical maps reflect the actual structure
of the DNA, its range as well as resolution can be adjusted within wide limits
by choosing the appropriate rare-cutting restriction enzymes and electrophoretic conditions. The combination of genetic and physical mapping
strategies has been essential for the isolation of genes responsible for
many human genetic disorders including cystic fibrosis and Huntington's
disease.

Long-range restriction maps are established from analyses of single and
double enzyme digests in much the same way as restriction analysis of small
DNA fragments, except hybridization rather than ethidium bromide staining
is used to visualize DNA fragments. The availability of a physical map not
only permits markers to be ordered more precisely than genetic linkage, but
also determines the size of the genomic region to search for candidate genes.
This will, in turn, influence the choice of cloning strategies used to isolate
genes of interest. As most of the restriction enzymes used in physical mapping
contain one or more CpG dinucleotides in their recognition sequence, potential sites of genes can be identified from the presence of CpG islands in the
physical map (4), since such islands have been associated with the 5′ end of
some genes. This chapter will describe the preparation of mammalian DNA
for PFGE, conditions for optimizing resolution in PFGE, construction of

long-range restriction maps, and potential problems associated with PFGE mapping.

2. Preparation of DNA for PFGE

Standard procedures for DNA preparation in solution do not yield intact high molecular weight DNA due to mechanical breakage and nuclease contamination during isolation. To avoid damage of large DNA by shear forces, two methods have been described for preparing high molecular weight DNA from cells embedded in agarose. One method routinely used is that described by Schwartz and Cantor, of embedding cells in agarose plugs (1). In this procedure, outlined in *Protocol 1*, cells resuspended with an equal volume of molten low melting temperature (LMT) agarose are aliquoted into chambers of a plastic mould to form plugs. Such moulds are usually supplied with the commercial PFGE apparatus, otherwise 1 ml syringes that are cut off at the front end and sealed with tape can be used to prepare samples. Once solidified, the plugs are treated with protease (e.g. Proteinase K) to digest cellular proteins, in the presence of detergent and high concentration of EDTA to inhibit nucleolytic activity. Cellular components released from this digestion are easily removed via diffusion during repeated washings of the plugs, leaving behind intact chromosomal DNA. The agarose matrix protects the embedded DNA from shear forces and provides an easy way to manipulate high molecular weight DNA. In the second method, cells are encapsulated as small agarose beads by pipetting the agarose–cell suspension into mineral oil while vortexing (5, 6). The size of the beads is controlled by vortexing speed and subsequent treatments are identical to those for agarose plugs.

Protocol 1. Preparation of human DNA in agarose plugs

Equipment and reagents

- 3.8% (w/v) tri-sodium citrate
- Lysis buffer: 155 mM NH_4Cl, 10 mM NH_4HCO_3, 0.1 mM EDTA, pH 7.4
- Phosphate-buffered saline (PBS): 144 mM NaCl, 10 mM KH_2PO_4 pH 7.4 with K_2HPO_4
- TE: 10 mM Tris–HCl, 1 mM EDTA, pH 7.4
- TE_{50}: 10 mM Tris–HCl, 50 mM EDTA, pH 7.4
- Cell lysis solution: 1% (w/v) N-lauroyl sarcosine, 0.5 M EDTA pH 8.0, 0.5 mg/ml Proteinase K (Boehringer Mannheim)
- 1.2% (w/v) LMT agarose (SeaPlaque GTG, FMC) in PBS
- 50 ml plastic tubes (Falcon) and 1.5 ml Eppendorf tubes

- PMSF (phenylmethylsulfonylfluoride): prepare fresh each time by dissolving 40 mg/ml PMSF in ethanol at room temperature. (**Caution:** PMSF is highly toxic and should be handled with gloves.)
- Perspex mould with slots of approximately 100 µl volume (available from BioRad and Pharmacia Biotech.)
- Haemocytometer
- Bench-top centrifuge
- Rocking platform
- Sterile medical gauze

Method

1. Collect cells in PBS from

> (a) **Blood**: mix 10 ml of blood with 1 ml of 3.8% tri-sodium citrate
>   (w/v) in a 50 ml plastic tube. Add 30 ml of lysis buffer and leave
>   on ice for 15 min, inverting the tube occasionally to mix the con-
>   tents. Pellet the cells by centrifugation at 1000 g for 10 min at room
>   temperature. Resuspend the cell pellet in 10 ml lysis buffer, leave
>   on ice for 10 min and centrifuge at 1000 g for 10 min at room
>   temperature.
>
> (b) **Tissue culture cells**: harvest cells by trypsinization using standard
>   procedures. Suspension cultures are centrifuged directly to collect
>   cells. Protocols for isolating cells from sperm and solid tissues are
>   described in Chapter 3, *Protocol 1*.
>
> **2.** Wash the cells twice in PBS and gently resuspend the pellet in 20 ml
>   PBS at room temperature.
>
> **3.** Count the cells using a haemocytometer and resuspend in PBS at a
>   concentration of 15×10^6 cells/ml.
>
> **4.** Mix the cells with an equal volume of 1.2% (w/v) LMT agarose pre-
>   pared in PBS and cooled to about 50°C in a water bath. To prevent
>   premature gelling, mix small volumes of cells and agarose in a 1.5 ml
>   Eppendorf tube.
>
> **5.** Dispense 100 μl of agarose–cell suspension mixture into chambers of
>   Perspex moulds, avoiding air bubbles.
>
> **6.** Allow the samples to solidify at 4°C or on ice for 10 min.
>
> **7.** Remove the samples from mould using a sterile inoculating loop.
>   Transfer between 30 and 40 plugs to 40 ml of cell lysis solution in a
>   50 ml plastic tube.
>
> **8.** Incubate the plugs at 50°C for 24–48 h on a rocking platform.
>
> **9.** Place the tubes on ice for 10 min to allow the plugs to firm up.
>
> **10.** Discard the cell lysis solution using a sterile medical gauze to retain
>   the plugs.
>
> **11.** Transfer the plugs to new 50 ml tubes and rinse four times with TE.
>
> **12.** Incubate the plugs in TE containing 40 μg/ml PMSF, at 50°C for
>   30 min, to inactivate the Proteinase K.
>
> **13.** Pour off the PMSF solution and wash the plugs in TE for 30 min on a
>   rocking platform. Repeat the washes at least three times.
>
> **14.** Store the plugs in TE$_{50}$ at 4°C.

Each new preparation of DNA samples should be tested for nuclease
contamination by fractionation on pulsed field (PF) gels. Intact DNA
samples should remain trapped in the wells on ethidium bromide-stained
gels.

3. Restriction enzyme digestion of agarose-embedded DNA

3.1 Choice of enzymes for long-range restriction mapping

The chromosomes of many microorganisms and some lower eukaryotes such as yeast and trypanosomes are small enough to be resolved as individual bands by PFGE. In contrast, most mammalian chromosomes are too large and require rare-cutting restriction enzymes to reduce DNA fragments to under 3 Mb. Fragments larger than this usually result from partial digestion. Rare-cutting restriction enzymes can be divided into two classes: those that have eight or more nucleotides in their recognition sequence and those that contain one or more CpG dinucleotides in their recognition site (*Table 1*). The latter class of restriction enzymes is sensitive to cytosine methylation and they tend to generate larger fragments because the CpG dinucleotide is rare in the mammalian genome, has a non-random pattern of distribution, and the majority of CpG sites are methylated. Unmethylated CpG sites tend to occur at high frequency in regions known as CpG islands. All housekeeping genes and some tissue-specific genes are associated with CpG islands at their 5' ends, thus localization of these islands by restriction enzyme mapping is a powerful means to identify expressed sequences (7). However, this clustering of CpG sites can also present significant problems if most rare-cutting enzymes cut around the same site, thereby reducing the chances of bridging neighbouring islands to produce a contiguous map. This disadvantage could be partially overcome by using enzymes that lack CpG in their recognition sequence, or to perform partial enzyme digestions (8). It is noteworthy that some rare-cutting restriction enzymes exhibit marked site preferences; New

Table 1. Rare-cutting restriction enzymes

Enzyme	Recognition site	Enzymes	Recognition site
*Asc*I	GG/CGCGCC	*Nru*I	TCG/CGA
*Bss*HII [a]	G/CGCGC	*Pac*I	TTAAT/TAA
*Cla*I	AT/CGAT	*Sac*II	CCGC/GG
*Eag*I	C/GGCCG	*Sal*I	G/TCGAC
*Fsp*I	TGC/GCA	*Sfi*I [a]	GGCC(N$_4$)/NGGCC
*Ksp*I [b]	CCGC/GG	*Sma*I	CCC/GGG
*Mlu*I	A/CGCGT	*Spl*I	C/GTACG
*Nae*I	GCC/GGC	*Srf*I	GCCC/GGGC
*Nar*I	GG/CGCC	*Xho*I	C/TCGAG
*Not*I	GC/GGCCGC		

[a] The temperature for optimum activity of these enzymes is 50°C.
[b] *Ksp*I is an isoschizomer of *Sac*II but is unaffected by CpG methylation.

England Biolabs observed differences in the rate of cleavage by a distinct group of enzymes (*Nar*I, *Nae*I, and *Sac*II) regardless of the amount of enzymes used or the length of the incubation period. The reasons for this phenomenon remain unknown. The choice of enzymes for physical mapping will depend on the genomic region being analysed, some regions (e.g. DMD region) have been found to be deficient for CpG islands (9) while other regions, such as the Wilms tumour locus (10), are rich in these sites.

The method recommended to digest high molecular weight genomic DNA embedded in agarose plugs with restriction enzymes is outlined in *Protocol 2*.

Protocol 2. Restriction enzyme digestion of DNA in agarose

An agarose plug prepared according to *Protocol 1* will contain about 7.5 $\times 10^5$ cells; roughly equivalent to about 5 µg of DNA. Use between one-third or one-half of a plug for each enzyme digest and include a negative control which contains all the reaction components except the restriction enzyme.

Equipment and reagents

- TE (see *Protocol 1*)
- TE$_{50}$ (see *Protocol 1*)
- Sterile scalpel
- Bovine serum albumin (BSA, 10 mg/ml)
- Spermidine trihydrochloride (100 mM, Sigma)

- 10 × Restriction enzyme buffer (provided by manufacturer)
- Restriction enzyme
- Sterile distilled water
- Incubator
- 1.5 ml plastic Eppendorf tubes

Method

1. Determine the number of plugs required for an experiment. Wash the plugs that were stored in TE$_{50}$ at least three times in 20–30 volumes of TE for 30 min each time, with gentle agitation.

2. Cut the agarose plugs with a sterile scalpel into thirds or halves.

3. Set up the following in a 1.5 ml Eppendorf tube:
Agarose plug (see *Protocol 1*)	30–50 µl
BSA (10 mg/ml)	2 µl
Spermidine (100 mM)	4 µl
Restriction enzyme buffer (10 ×)	20 µl
Restriction enzyme	10–20 units
Sterile distilled water	to a final volume of 200 µl

4. Leave on ice for 30 min for the agarose plug to equilibrate with the buffer and restriction enzyme.

5. Digest samples for 8 h to overnight at the appropriate temperature. Add a second aliquot of enzyme midway through the incubation period.

Protocol 2. *Continued*

6. Chill the tubes on ice for 10 min and load the plugs directly on to the gel or store them in TE$_{50}$ at 4°C.

7. For double restriction enzyme digestions, set up digests for the restriction enzyme with a lower salt requirement first. At the end of the first enzyme digest, aspirate the reaction buffer and rinse the plug in TE. Leave on ice for 10 min and repeat the washing. Remove the TE, add fresh reaction buffer and the second restriction enzyme.

3.2 Partial restriction enzyme digests

Partial enzyme digests are particularly useful in long-range restriction mapping to confirm linkage between DNA markers and to extend the size of the restriction map generated around any one marker or to map across CpG islands (*Figure 6* and Section 5.2). Some rare-cutting enzymes (e.g. *Sal*I) naturally generate partial digests regardless of experimental conditions.

Partial digestion of DNA embedded in agarose can be produced using standard procedures of limiting either enzyme concentration, or reaction time, or magnesium concentration (11), or by the simultaneous addition of methylase and restriction enzyme to compete for the same recognition site (12). Alternatively, Weil and McClelland (13) have shown that the combination of site-specific methylases and restriction enzymes containing partially overlapping recognition sequences can be used to restrict large DNA in a limited and predictable manner. A recent report described the use of *Cla*I, an adenosine methylase, in conjunction with *Dpn*I, a methyl-adenosine dependent restriction endonuclease, to create ultra-rare restriction sites in eukaryotic chromosomes of *Saccharomyces cerevisiae* and *Schizosaccharomyces pombe* (14). The advantage of this strategy is that adenosine methylation does not occur naturally in eukaryotic genomes, hence there is no pre-existing methylation to interfere with the restriction analysis and therefore it does not rely on competing reactions between methylase and restriction enzyme to produce large DNA fragments. By using different adenosine methylases whose recognition sequences partially overlap with that of *Dpn*I, this approach can be applied to restrict large mammalian chromosomes in a limited and controlled manner.

4. Electrophoresis

4.1 General considerations

All PFGE described in this chapter were performed using the CHEF-DRII system (BioRad). The conditions and parameters described here can be applied, with minor modifications, to other CHEF apparatus such as Pulsaphor™ (Pharmacia) and Rotaphor™ (Biometra). During PFGE, DNA molecules undergo continuous reorientation caused by periodic changes in alternating electric fields. The range over which DNA molecules can be separated is a function of several parameters. These include the pulse time, pulse angle, electrophoresis run time, voltage, agarose concentration, buffer concentration, and temperature. A detailed discussion of the effects of these parameters on the resolution and mobility of DNA molecules in PFGE can be found in Chapter 1 and elsewhere (15). For most applications, changing the pulse time is usually sufficient to achieve the intended separation or resolution.

To separate DNA molecules greater than 2 Mb, a lower agarose concentration, longer pulse times, and reduced voltage are required to prevent breakage and trapping of the large DNA molecules in the agarose matrix. Decreasing the pulse angle from 120° to 106° has been found to increase mobility of mega-sized DNA molecules, thereby reducing the overall electrophoretic run times (15). However, adjustment of pulse angle is only possible in selected apparatus such as CHEF-DRIII and CHEF Mapper (BioRad).

The ability to detect and accurately size bands of interest on pulsed field gel blots is dependent not only on choosing the appropriate conditions to optimize resolution but is also influenced by the amount of DNA loaded in the gel. Pulsed field gels are especially sensitive to sample overload. As the amount of DNA increases, its rate of migration is retarded correspondingly, resulting in larger than true sizes. This is shown in *Figure 1* where migration of bands in lane 1 are retarded relative to lane 2. Similarly in *Figure 8a*, the specific hybridizing band from lane C is shifted higher relative to lane JD, due to increased amounts of DNA present in that lane; as judged from the ethidium bromide stained gel and intensity of the hybridizing band.

4.2 Size markers

DNA size standards are particularly important in PFGE analysis to

(a) determine the sizes of unknown DNA fragments after hybridization;
(b) assess the effectiveness of the chosen electrophoretic programme for the intended separation or resolution.

Changes in pulse times can have a dramatic effect on the size range over which DNA molecules can be separated as well as its 'window' of resolution; a broad size range separation is usually achieved at the expense of resolution, resulting in less accurate size determination (see Section 4.3).

Since the introduction of PFGE, the steady increase in the number of DNA markers for different size ranges has enabled a more accurate assessment of sizes for PFGE analysis. *Table 2* lists some of the commonly used size markers that are available commercially. When using markers that have a constant increment in size (e.g. lambda oligomer or MidRange II PFG marker), it is advisable to include a second marker such as *S. cerevisiae* to serve as a reference marker, in case the first band of the ladder has run off the gel.

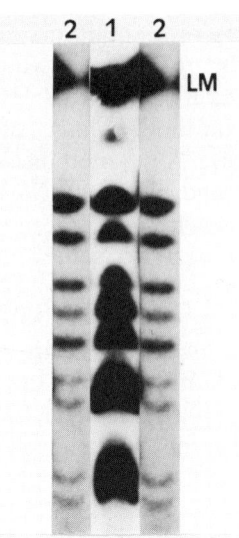

Figure 1. Effects of sample overloading in pulsed field gels. Genomic DNA isolated from blood was digested with *Sal*I and subject to electrophoresis in a 1% (w/v) agarose gel. About 10 µg and 3 µg of DNA were loaded in lanes 1 and 2, respectively. Hybridization of an anonymous probe to the Southern blot of the gel shows that the rate of DNA migration in lane 1 is retarded relative to lane 2 and hybridization signals were more diffuse due to increased amounts of DNA. LM refers to limiting mobility.

Table 2. Commonly used size markers for PFGE

Marker	Size range (kb)	Supplier
High molecular weight	8.2–48.5	BRL
MidRange II PFG	DNA ladder from 24–600	New England Biolabs (NEB)
Lambda oligomer	DNA ladder from 48.5–800	BioRad, NEB, etc.
S. cerevisiae chromosomes	220–2200	BioRad, NEB, etc.
Hansenula wingei chromosomes	1000–3300	
Candida albicans chromosomes	1000–3000	Clontech
Schizosaccharomyces pombe chromosomes	3500–5700	BioRad, etc.

4.3 Optimizing resolution in PF gels

Resolution in PF gels is usually assessed from the separation of size markers after gel electrophoresis. Lerman and Sinha (16) proposed that the distance between bands of DNA marker and its band width be used as a measure of resolution in gel electrophoresis; thus a separation of 1 cm with a band width of 1 mm is more useful than a separation of 2 cm with a band width of 5 mm

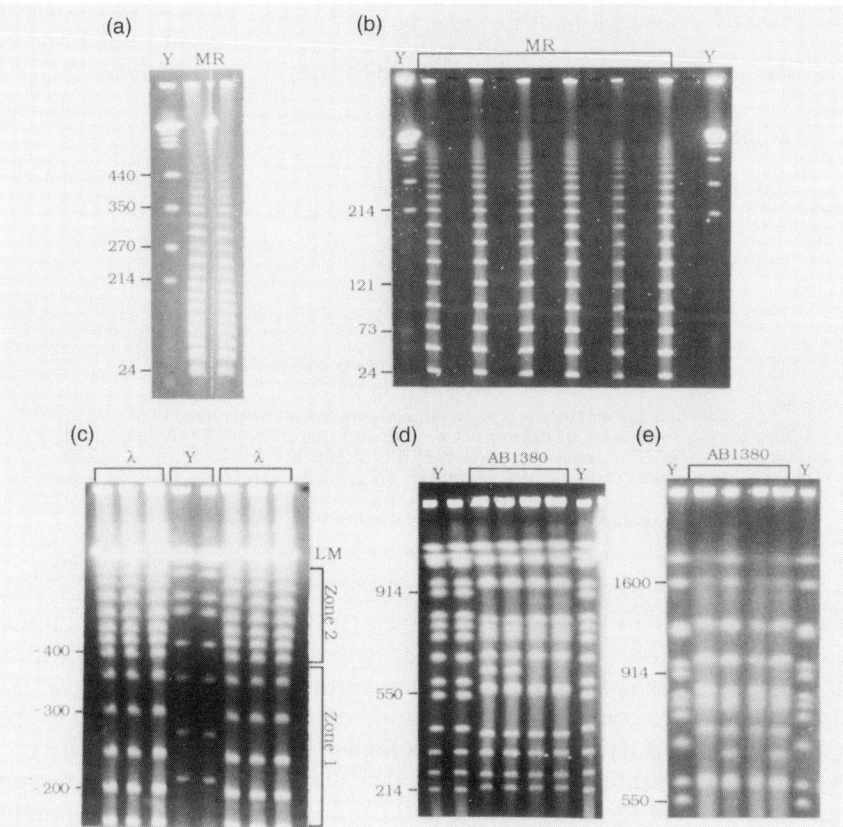

Figure 2. Multiple zones of resolution in pulsed field gels. (a) and (b) show the separation of MidRange II PFG marker (MR) under different electrophoretic conditions. The pulse times used in (a) were initially 25 sec for 18 h followed by 45 sec for 18 h, and in (b) a pulse time ramp from 5 sec to 25 sec for 32 h. (c) The separation of lambda oligomer DNA demonstrates that separation in pulsed field gels is not linear and resolution within the same gel can be divided into windows or zones as shown. Highest resolution is obtained in the region where oligomer bands of marker DNA are linearly separated and widely spaced with minimum band width. LM refers to limiting mobility. (d) and (e) show the separation of yeast chromosomes of *S. cerevisiae* strains YNN295 (lane Y) and AB1380 (containing a YAC) using different electrophoretic conditions as described in *Table 3*. By choosing the appropriate conditions, different size ranges can be optimized to facilitate size determination of unknown fragments.

(compare *Figure 2a* and *b*). Oligomers (e.g. lambda or MidRange II PFG marker) that are linearly separated offer high resolution and this enhances size estimations. In a typical PF gel, however, multiple 'zones' or 'windows' of resolution can be identified in different regions of the same gel (17), this is reflected by the uneven spacing of oligomer marker DNA. In *Figure 2c*,

Table 3. Conditions for PFGE separations [a]

Range (kb)	Voltage (V/cm)	Pulse time (sec)	Duration (h)
24–300	5.6	Ramp from 5 to 25	30
24–600	5.6	25 [b]	20
		50	18
100–1000	5.6	60 [b]	18
		100	13
		150	8
220–1600	5.6	100 [b]	24
		180	24
1000–3000	4.5	120 [b]	24
		240	36
3500–5700	1.8	Ramp from 300 to 1800	96

[a] This table lists the electrophoretic conditions that are routinely used in our laboratory for the separation of different size ranges of DNA. All PFGE were performed with the CHEF-DRII apparatus at 12°C in 0.5 × TBE buffer. A 1% (w/v) agarose concentration was used for DNA separation up to 1600 kb and 0.6% for sizes larger than that.
[b] Refers to the use of multiple pulsed time phases.

separation in this PF gel can be conveniently divided into two zones of resolution; bands of λ DNA in the lower half of the gel (between 150 kb and 350 kb) are evenly spaced and separated further than those in the 400–700 kb range which are compressed. Estimation of sizes from zone 1 will be more accurate compared to zone 2 due to its higher resolution.

Linearity of separation can be improved by the use of (a) pulse time ramping, which involves a progressive change in the pulse time interval throughout the duration of gel electrophoresis or (b) performing multiple pulse time phases sequentially within a gel run (see *Table 3*). Hence, by choosing the appropriate electrophoretic conditions, different windows of resolution can be enlarged to provide a more accurate assessment of DNA sizes. This is best illustrated in *Figure 2a* and *2b* which shows the separation of the MidRange II PFG marker under different electrophoretic conditions. Conditions for electrophoretic separation in various size ranges are shown in *Table 3*.

4.4 Gel preparation

Gels are prepared using the casting stand supplied with the commercial gel boxes and is outlined in *Protocol 3*.

Protocol 3. Casting and loading the gel

Equipment and reagents

- Agarose (SeaKem, FMC)
- 0.5 × TBE: 45 mM Tris, 45 mM boric acid, 0.5 mM EDTA, pH 8.3
- Gel casting stand and comb
- Plastic inoculating loop and sterile scalpel
- PFGE apparatus

Method

1. Prepare 1% (w/v) agarose in 0.5 × TBE. To separate DNA greater than 2 Mb, use between 0.5% and 0.8% (w/v) agarose.

2. Assemble the gel casting stand and comb. Ensure that it is resting on a level surface.

3. Pour agarose and leave to solidify.

4. There are two methods of loading agarose plugs in gels.

    (a) Use an inoculating loop to remove plug from Eppendorf tube and place it on a sterile scalpel blade. Remove excess buffer or TE_{50} with tissue paper and gently push plug into gel slot with inoculating loop. Seal wells with agarose.

    (b) Alternatively, agarose plugs are cast *in situ* with the gel. Place plugs on the teeth of a gel comb placed horizontally, remove excess buffer or TE_{50} with tissue paper, and allow to air dry slightly so that plugs stick to the comb. Assemble casting stand by placing the gel comb vertically and pour the agarose to cast the gel. Optional: After removing the comb once the gel is set, pipette more agarose into the holes left from the comb behind the plugs. Avoid trapping any air bubbles. Otherwise, fill the holes with running buffer.

5. Transfer the gel to the PFGE apparatus electrophoresis chamber and add running buffer (0.5 × TBE) to cover the gel. Start electrophoresis using appropriate parameters.

4.5 Southern blot transfer of PF gels

Large DNA molecules fractionated on PF gels must be partially hydrolysed for efficient transfer on to membranes. This is accomplished by either acid depurination or UV irradiation followed by standard blotting procedures as outlined in *Protocol 4*.

Protocol 4. Staining and blotting of PF gels

Equipment and reagents

- 0.25 M HCl
- Denaturing solution: 0.5 M NaOH, 1.5 M NaCl
- SSC (20 × stock solution): 3.3 M NaCl, 0.3 M Na_3-citrate
- Ethidium bromide (0.5 μg/ml) in 0.5 × TBE (see *Protocol 3*)

- Sterile distilled water
- Short-wave UV light transilluminator
- Nylon transfer membrane (Hybond N+, Amersham)
- UV light cross-linker (Stratalinker, Stratagene)

Method

1. Stain the gel with 0.5 μg/ml ethidium bromide in electrophoresis buffer for 30 min.

31

Protocol 4. *Continued*

2. Destain the gel for 30 min in electrophoresis buffer or distilled water and photograph the gel with a ruler by the side.

3. Place the gel in 0.25 M HCl and shake gently at room temperature for 8 min. Alternatively, UV irradiate the gel for 40–60 sec at 254 nm on a transilluminator.

4. Rinse the gel with distilled water and soak it in denaturing solution for 30 min with gentle shaking.

5. Invert the gel and set up alkaline capillary transfer to a nylon membrane (Hybond N+). Leave for 16–24 h.

6. Remove the membrane and rinse briefly in 2 × SSC.

7. Air dry the membrane and UV cross-link it in Stratalinker to immobilize the DNA. There is no need to bake positively charged membranes.

Figure 3 shows typical ethidium bromide-stained gels after PFGE. Radioactive probing of the Southern blot from gel *a* is shown in *Figure 6*. The asterisk in gel *b* highlights partial nucleolytic degradation in that sample.

4.6 Hybridization of PFGE blots

Standard procedures are used for hybridizing PFGE blots in either formamide or non-formamide based buffers as outlined in *Protocol 5*. Radioactively labelled probes should be tested for the presence of repetitive sequences prior to hybridization to PFGE blots, since such sequences can be effectively blocked by prehybridization with sonicated human placental DNA (Sigma). A simple test for the presence of repeat elements is to hybridize non-competed probes on panels or strips of human genomic DNA digested with frequently cutting restriction enzyme. *Figure 4* shows the hybridization of a pre-competed radioactively labelled whole cosmid on a PFGE blot. Label probes according to the random hexanucleotide priming method (18).

Protocol 5. Hybridization of PFGE blots

Equipment and reagents

- Hybridization buffer: 50% (v/v) formamide, 4 × SSC (see *Protocol 4*), 50 mM sodium phosphate buffer pH 7.2, 1 mM EDTA, 10% (w/v) dextran sulphate (Pharmacia), 1% (w/v) SDS, 50 μg/ml alkali-denatured salmon sperm DNA (Sigma), and 10 × Denhardts solution
- 100 × Denhardts solution: 2% (w/v) BSA Fraction V (Sigma), 2% (w/v) Ficoll 400 (Sigma), and 2% (w/v) polyvinylpyrrollidone (Sigma)
- Random hexamer priming labelling kit
- Ethanol

- TE (see *Protocol 1*)
- Sodium phosphate buffer pH 7.2: 1 M $Na_2HPO_4.7H_2O$, pH with H_3PO_4
- Strip solution: 0.2 × TE, 0.5% (w/v) SDS
- Sonicated human placental DNA (Sigma)
- Incubator
- Geiger counter
- X-ray film (Kodak or Fuji)
- Cling film
- X-ray cassette with intensifying screen
- −70°C freezer

Method

1. Prehybridize the PFGE blots in hybridization buffer for at least 6 h at 42°C.

2. Radioactively label the DNA probe according to the random priming method. Remove unincorporated labelled nucleotides by ethanol precipitation and resuspend the probe in 100 μl of TE. Denature the probe by heating to 95–100°C for 8 min and chill on ice.

3. For DNA probes containing repetitive sequences, resuspend the labelled probe in 100 μl TE after ethanol precipitation (from step 2) and add 100–200 μg of sonicated human placental DNA (Sigma) to a final volume of 176 μl. Boil for 8 min and chill on ice. Add 24 μl of sodium phosphate buffer and incubate at 65°C for 90 min. Use the probe directly without further treatment.

4. Pour off the prehybridization buffer and add 1×10^6 c.p.m./ml of denatured radioactive probe to the PF blots. Incubate at 42°C for about 16 h.

5. After hybridization, wash the membranes at 65°C in
   - $2 \times$ SSC, 0.1% (w/v) SDS for 30 min
   - $1 \times$ SSC, 0.1% (w/v) SDS for 30 min
   - $0.5 \times$ SSC, 0.1% (w/v) SDS for 20 min
   - $0.2 \times$ SSC, 0.1% (w/v) SDS for 20 min

   Monitor radioactivity on the PF blots at the end of each wash with a Geiger counter before proceeding to a higher stringency wash.

6. Wrap the membranes in cling film and expose to X-ray film in an X-ray cassette with intensifying screens at −70°C for 1–4 days.

7. To remove the old signals on PF blots before rehybridization with a new probe, preheat the strip solution to 80°C and add directly to the membranes. Leave at 65°C for 30 min with gentle agitation and check residual activity by autoradiography.

5. Construction of maps from hybridization results

5.1 Strategies for long-range restriction mapping

An important element in the construction of long-range restriction maps is the availability of markers or probes. The markers used for PFGE mapping are usually single copy genomic DNA fragment or cDNA clones, as these lack repeat sequences and are easy to use. However, whole cosmids (see *Figure 4*), rare-cutting restriction enzyme linking clones (19, 20), and end clones of specific fragments isolated by preparative PFGE (21, 22) have also

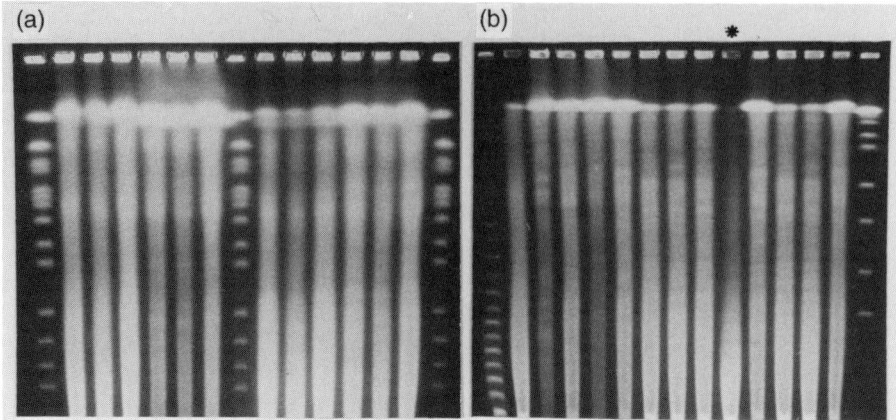

Figure 3. Ethidium bromide-stained pulsed field gels containing human genomic DNA isolated from blood and digested with various rare-cutting restriction enzymes. Electrophoresis was performed at 5.6 V/cm, the pulse times in (a) were initially 60 sec for 21 h followed by 100 sec for 15 h and 150 sec for 8 h; and in (b) 25 sec for 20 h followed by 50 sec for 18 h. Chromosomes prepared from *S. cerevisiae* (BioRad) and MidRange II PFG marker (NEB) were used as size standards. Results of radioactive hybridization of the Southern blot from gel A is shown in *Figure 6*. The asterisk in gel B highlights partial DNA degradation.

been used in map construction. The endeavour will be further assisted if the relative positions of markers are known. This information can be derived from *in situ* hybridization (23), genetic linkage or radiation hybrid mapping (24), or from chromosomal rearrangements (e.g. deletion or translocation). However, if markers are not available or separated at great distances from each other, cloning the region of interest and partial mapping in YACs would be the alternative and this is discussed in Chapters 5 and 6.

Sources of DNA for physical mapping include blood, sperm, somatic cell hybrids, and lymphoblastoid cell lines. The degree and pattern of DNA methylation are highly variable amongst these different cell types and at different developmental stages, hence careful consideration should be given to the source of DNA used, since most of the rare-cutting restriction enzymes contain one or more CpG dinucleotides in their recognition sequence and are sensitive to cytosine methylation. When comparative mapping between wild-type and mutant DNA is performed (as in the detection of chromosomal rearrangements as described in Chapter 3), it is advisable to use the same tissue source to minimize methylation differences.

To construct a meaningful long-range restriction map, it is important to establish linkage between neighbouring markers. This entails sequential hybridization of the same PFGE blot to different markers to assess linkage. As a first step, prepare PFGE blots consisting of single restriction enzyme

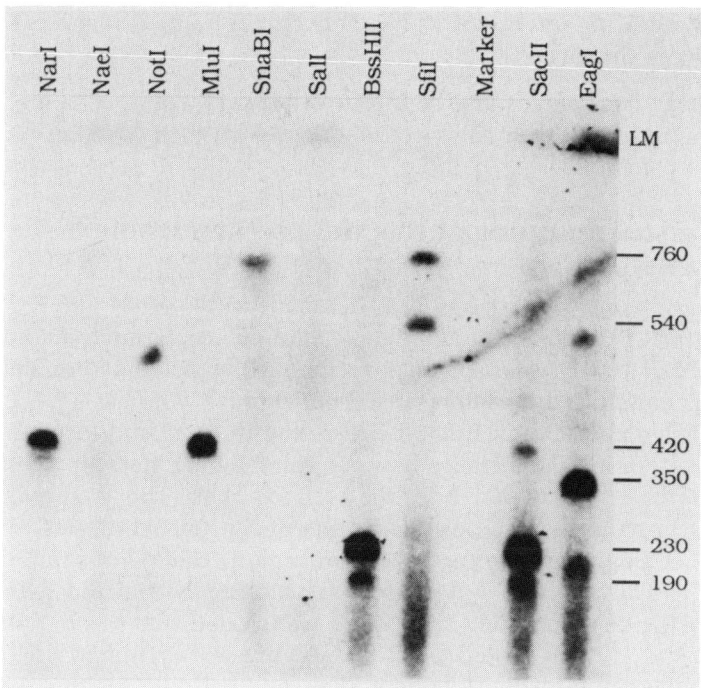

Figure 4. Hybridization of a radioactively labelled cosmid on a PFGE blot. Uncut radio-labelled cosmid DNA was pre-competed with sonicated human placental DNA according to *Protocol 5* and hybridized to a PFGE Southern blot containing human genomic DNA isolated from blood. The electrophoretic conditions used were as described in *Figure 3a*. LM refers to limiting mobility.

digests separated in different size ranges, for example: (a) 24–600 kb, (b) 220–1600 kb, and (c) 1000–3000 kb. Sequential hybridization of markers to these blots not only assesses possible linkage but will also reveal if the region under study is rich in CpG islands. Align autoradiographs to determine if there is a common pattern or band from different probe hybridizations. If two probes hybridized to a 'common' band, the result is only suggestive of linkage and further confirmation is required. This is because in CpG rich regions, digestion with enzymes such as *Bss*HII, *Eag*I, and *Sac*II often result in small fragments of similar sizes. To exclude random co-migration of restriction fragments, ensure that;

(a) both markers also hybridized to a common band in other restriction enzyme digests or to the largest fragment in partial restriction enzyme digests;

(b) in double restriction enzyme digests, the sum of the small hybridizing fragments of each marker should add up to the size of the band detected in the single restriction enzyme digests (see Section 5.2);

(c) the hybridizing intensities of bands detected by both markers should be relatively similar.

Once linkage has been established for two markers, construct a more refined restriction map from the analysis of double digests with different rare-cutting restriction enzymes.

5.2 Physical mapping of the McLeod syndrome locus as an example

The McLeod syndrome gene locus (XK) maps within Xp21 and is flanked by probe 3BH/R 0.3 (DXS709) and the gene for chronic granulomatous disease (CGD). 3BH/R 0.3 was isolated by cloning the deletion breakpoint of a Duchenne muscular dystrophy (DMD) patient, JD, who has a 6 Mb deletion in Xp21 (*Figure 5*). Physical mapping around the McLeod locus was undertaken to determine the distance between the flanking markers and therefore the size of the region to search for the McLeod gene. The 'framework' of the map was constructed based on the results of *Sfi*I partial digests, since most of the other single enzyme digests in this region resulted in small fragments (*Figure 6*). The order of *Sfi*I sites was established from the hybridization pattern of the distal marker, 3BH/R 0.3, which detected common fragments of 760 kb and 1120 kb (as did CGD-cDNA) but also identified a 540 kb band as the smallest hybridizing fragment. This finding indicates that 3BH/R 0.3 resides on a 540 kb *Sfi*I fragment adjacent and distal to the 220 kb *Sfi*I fragment on which CGD lies, hence both probes are linked on a common

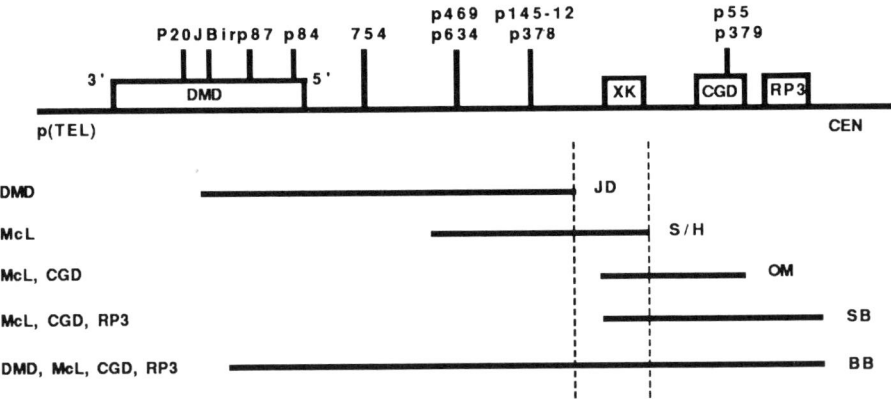

Figure 5. Schematic representation of the Xp21 region. Map of Xp21 showing the relative positions of genes and cloned loci. Below the map, the approximate sites and extent of deletions in different patients are indicated by a thick bar. The clinical manifestations of patients are shown on the left. McLeod locus (XK) was mapped, by deletion analysis, to a region between the proximal deletion endpoints of patients JD and S/H as shown by the vertical broken lines. (Reprinted and modified with permission from ref. 25.)

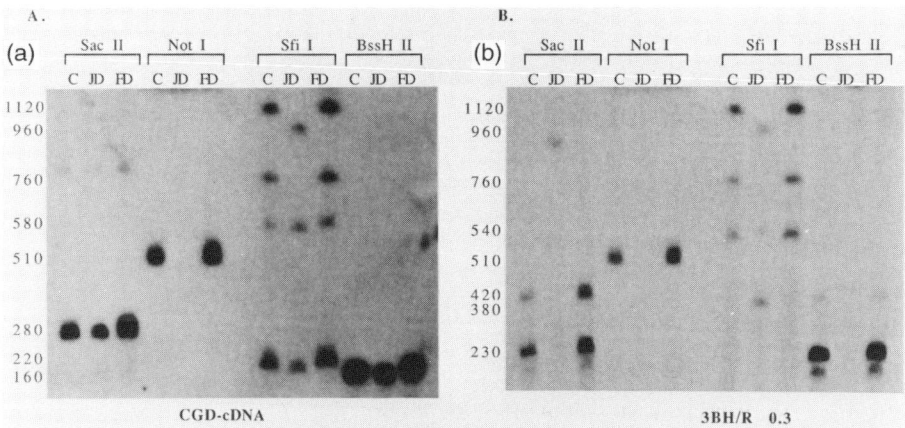

Figure 6. Physical linkage of 3BH/R 0.3 and CGD-cDNA to common *Sfi*I and *Not*I fragments. Flanking markers 3BH/R 0.3 and CGD-cDNA were hybridized sequentially to a PFGE blot containing genomic DNA from a healthy male individual, C; a DMD patient, JD; and a McLeod patient, FD. The DNA was digested to completion with the restriction enzymes indicated and electrophoresis was performed as described in *Figure 3a*. Both markers co-recognized common 760 kb and 1120 kb *Sfi*I fragments as well as a 510 kb *Not*I fragment on non-deleted X chromosome, while aberrant size bands were detected in patient JD. (Reprinted with permission from ref. 25.)

760 kb *Sfi*I partial digestion product. Further evidence of linkage was obtained when both flanking markers co-recognized a 510 kb *Not*I fragment (*Figure 6*). Coincidental sizes and co-migration of bands were excluded based on the analysis of double digests. For example, in double digests of *Sfi*I/*Not*I, CGD-cDNA hybridized to partial digestion products of 160 kb and 510 kb while 3BH/R 0.3 detected bands of 350 kb and 510 kb; the sum of the two smallest hybridizing bands is equivalent to the 510 kb band observed in single *Not*I digests (*Figure 7*). Further analysis with several other rare-cutting restriction enzymes led to the construction of a refined map around the McLeod locus (*Figure 9*). *Figure 8* shows PFGE analysis of normal X chromosome and the deleted X chromosome of patient JD around the McLeod locus. A detailed discussion on the physical mapping of the McLeod locus can be found elsewhere (25).

5.3 Potential problems

5.3.1 Cross-reactivity of probes

Figures 7b and *8b* show the hybridization of probe 3BH/R 0.3 which detects strongly hybridizing bands as well as a fainter 190 kb band. The latter was established to be a result of cross-hybridization to the region between CpG islands (I) and (II) (see reference 25 for discussion). Such cross-hybridizing bands can be confusing especially if it co-migrates with specific restriction

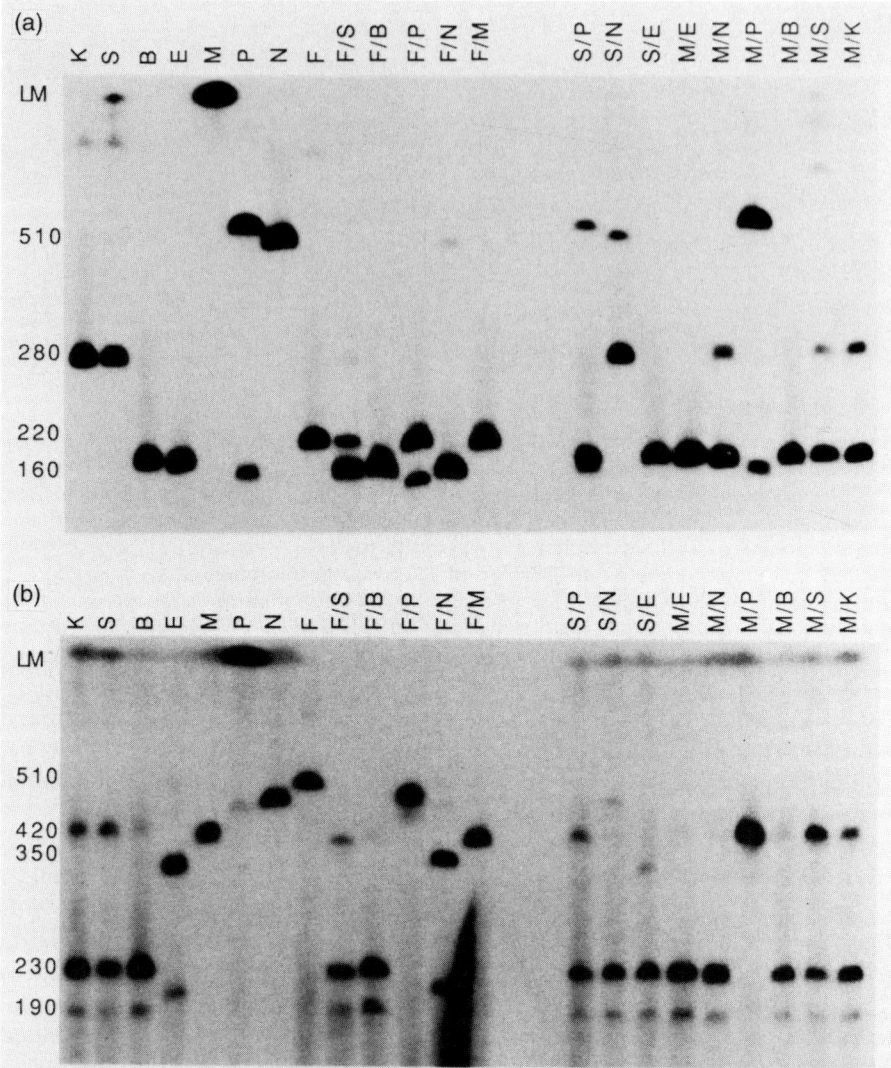

Figure 7. PFGE analysis of the McLeod locus. Sequential hybridization of (a) CGD-cDNA and (b) 3BH/R 0.3 to a Southern blot containing genomic DNA from a healthy male individual. The DNA was digested to completion with the restriction enzymes indicated and resolved using pulse times of 25 sec for 17 h followed by 60 sec for 16 h and 120 sec for 5 h. Restriction enzymes used were *Ksp*I (K), *Sac*II (S), *Bss*HII (B), *Eag*I (E), *Mlu*I (M), *Fsp*I (P), *Not*I (N), and *Sfi*I (F). LM refers to limiting mobility. The fainter hybridizing bands of 190 kb and 210 kb (for *Eag*I and *Sfi*I/*Not*I digests) was detected by 3BH/R 0.3 as a result of cross-hybridization. (Reprinted with permission from ref. 25.)

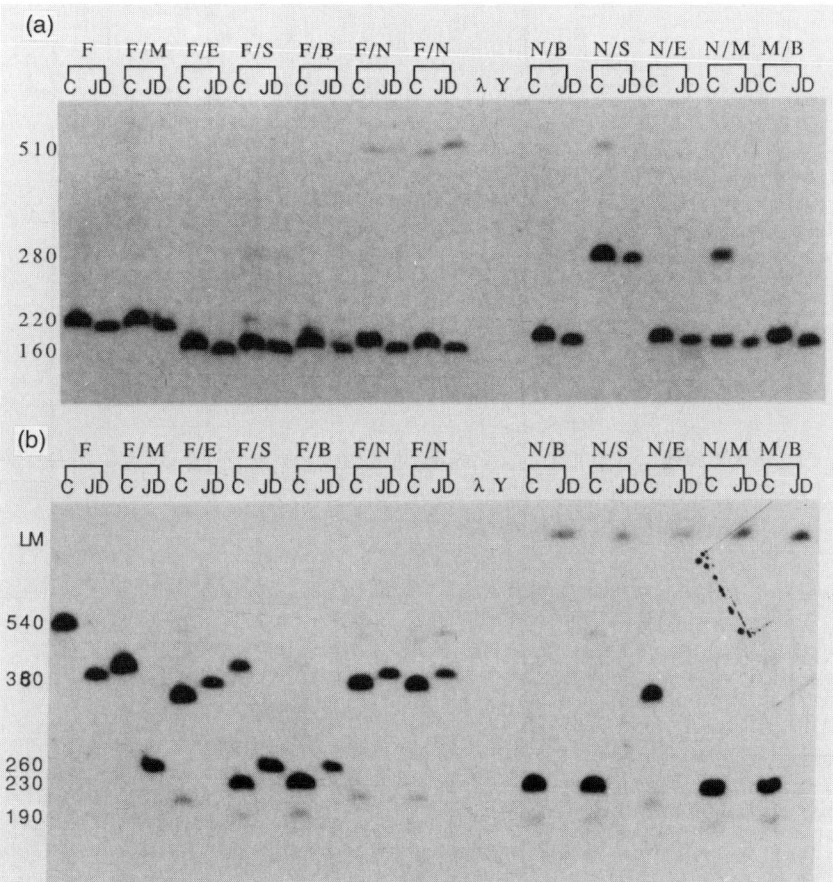

Figure 8. PFGE analysis around the McLeod locus in non-deleted X chromosome and deleted X chromosome of patient JD. (a) Hybridization of CGD-cDNA to PFGE blot containing DNA from a healthy male individual (lane C) and patient JD. Electrophoresis was performed as described in *Figure 7*. No aberrant size fragments were detected between DNA of patient JD and control around the CGD locus. However, migration differences in identical enzyme digests between DNA of control and of JD were consistently observed. This is due to higher DNA concentration used in the control sample. (b) Hybridization to the same blot with 3BH/R 0.3, which detected altered size fragments in JD for all enzymes tested. Size markers were lambda oligomers and chromosomes from *S. cerevisiae*. Restriction enzymes are abbreviated as in *Figure 7*, LM refers to limiting mobility. (Reprinted with permission from ref. 25.)

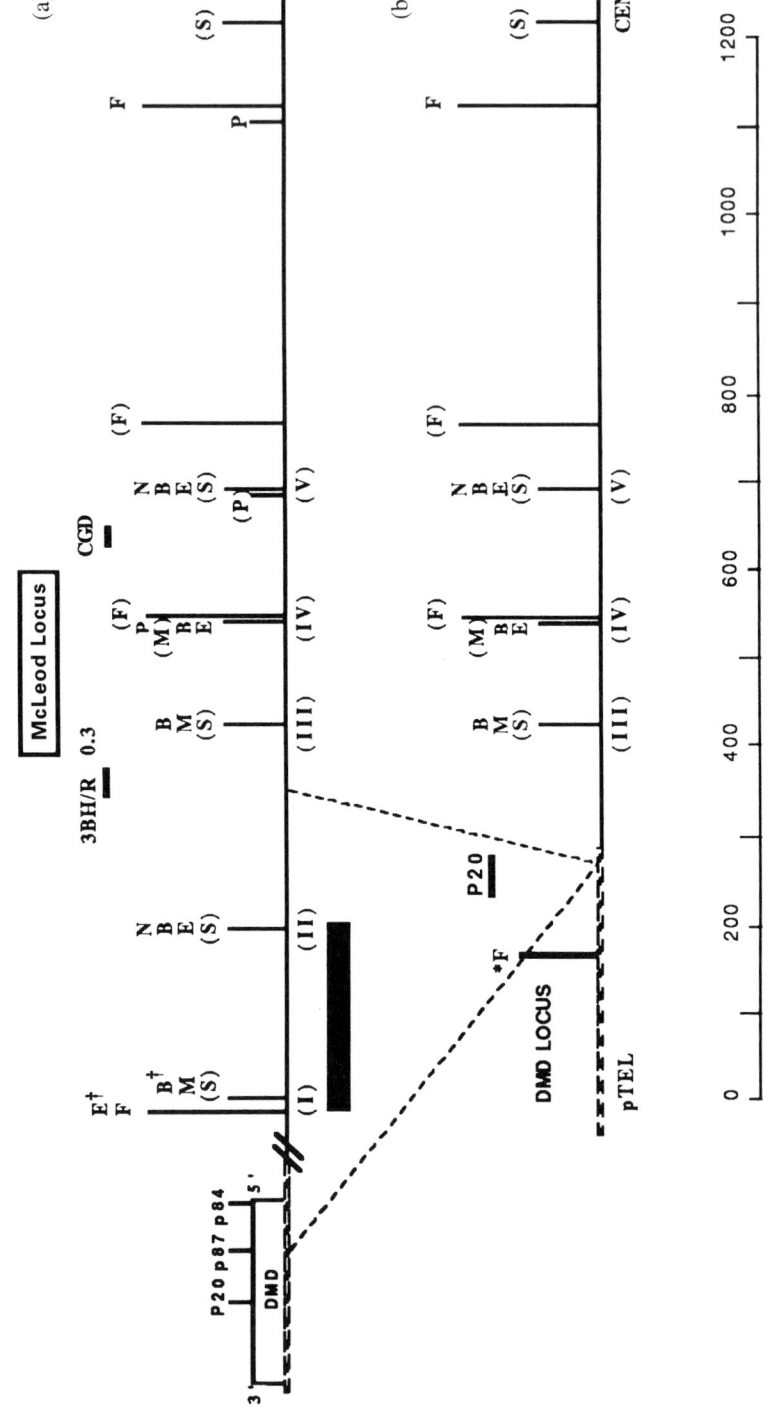

Figure 9. Comparative physical maps of the region around the McLeod locus in normal and in the deleted X chromosome of patient JD. (a) Restriction map encompassing McLeod and CGD loci which was constructed using DNA isolated from blood of a healthy male individual and digestion with all combinations of restriction enzymes *Ksp*I (K), *Sac*II (S), *Bss*HII (B), *Eag*I (E), *Mlu*I (M), *Fsp*I (P), *Not*I (N), and *Sfi*I (F). CpG islands identified by the presence of rare-cutting restriction enzymes sites have been designated I–V. Comparative mapping against the deleted X chromosome of patient JD has enabled the assignment of the new marker, 3BH/R 0.3, to a region located 30–150 kb distal to island III. The minimum regions to which CGD, P20, and the McLeod locus could be assigned are indicated above the map. The black bar below islands I and II highlights the region of cross-hybridization detected by 3BH/R 0.3; the presence of *Bss*HII and *Eag*I sites at island I (indicated by [†]) was inferred from the size of the cross-hybridizing band observed. (b) Restriction map of the deleted X chromosome in patient JD. The broken lines joining (a) and (b) indicate a region of about 6000 kb that was deleted in JD, and the horizontal hatched lines denote sequences from the DMD locus. Partially cleaved restriction enzyme sites are shown in parentheses. (Reprinted with permission from ref. 25.)

fragments. However, careful analysis and its weaker hybridization intensities should prevent errors in interpretation.

If cDNA probes are used for mapping in somatic cell hybrids, it is important to ensure that there is no cross-hybridization of cDNA clones to hamster or rodent DNA.

5.3.2 Partial digestion of DNA

Incomplete digestion of DNA may result from partial methylation or site preferences of the restriction enzyme. Partial digests may also occur naturally for certain restriction enzymes such as *Sal*I. The advantages of partial digests in map construction have been discussed earlier (see p. 26). However, when three or more fragments result from a single enzyme digest (as in *Sal*I digests shown in *Figure 1*), analysis of double digests will be extremely complex.

5.3.3 Genomic organization and distribution of CpG islands

The mapping of human disease genes in different regions of the genome reveals that unmethylated CpG dinucleotides are not randomly distributed but are clustered to form 'CpG islands'. These islands have been shown to be associated with the 5′ ends of some genes and regions rich in CpG islands appear to correspond to the negatively staining Giemsa or R bands (26). Physical mapping in CpG-rich regions often results in small restriction fragments, which limits the range of the map. On the other hand, mapping in CpG-poor regions will have a greater range because of the larger fragments generated, but suffers from lower resolution.

Pulsed field gel electrophoresis has enabled large regions of the genome to be analysed without cloning. This has provided insights on the structure and organization of complex genomes. Physical maps generated from PFGE analysis have been instrumental in positional cloning of human genes and has facilitated the diagnosis of chromosomal abnormalities.

Acknowledgements

We thank Hans Lehrach for advice and discussions on PFGE mapping, Iona Millwood for assistance in the preparation of this manuscript, and members of the laboratory for helpful comments on this manuscript. This work was supported by the Imperial Cancer Research Fund.

References

1. Schwartz, D. C. and Cantor, C. R. (1984). *Cell*, **37**, 67.
2. Carle, G. F. and Olson, M. V. (1984). *Nucleic Acids Res.*, **12**, 5647.
3. Chu, G., Vollrath, D., and Davis, R. W. (1986). *Science*, **224**, 1582.
4. Estivill, X., Farall, M., Scambler, P. J., Bell, G. M., Hawley, K. M., Lench, N. J., *et al.* (1987). *Nature*, **326**, 840.
5. Cook, P. R. (1984). *EMBO J.*, **3**, 1837.
6. Overhauser, J. (1992). In *Methods in molecular biology* (ed. M. Burmeister and L. Ulanovsky), Vol. 12, pp. 129–34. Humana Press, New Jersey.
7. Lindsay, S. and Bird, A. P. (1987). *Nature*, **321**, 336.
8. Barlow, D. P. (1992). In *Methods in molecular biology* (ed. M. Burmeister and L. Ulanovsky). Vol. 12, pp. 107–28. Humana Press, New Jersey.
9. Burmeister, M., Monaco, A. P., Gillard, E. F., van Ommen, G. B. J., Affara, N. A., Ferguson-Smith, M. A., *et al.* (1988). *Genomics*, **2**, 189.
10. Bonetta, L., Kuehn, S. E., Huang, A., Law, D. J., Kalikin, L. M., Koi, M., *et al.* (1990). *Science*, **250**, 994.
11. Hoheisel, J. D., Nizetic, D., and Lehrach, H. (1989). *Nucleic Acids Res.*, **17**, 9571.
12. Hanish, J. and McClelland, M. (1989). *Anal. Biochem.*, **179**, 357.
13. Weil, M. D. and McClelland, M. (1989). *Proc. Natl. Acad. Sci. USA*, **86**, 51.
14. Wilson, W. W., Mebane, E. W., and Hoffman, R. M. (1993). *Anticancer Res.* **13**, 17.
15. Birren, B. and Lai, E. (1993). In *Pulsed field gel electrophoresis: A practical guide*, pp. 129–40. Academic Press, San Diego.
16. Lerman, L. S. and Sinha, D. (1990). In *Electrophoresis of large DNA molecules. Theory and application. Current communications in cell and molecular biology* (ed. E. Lai and B. W. Birren), Vol. 1, pp. 1–8. Cold Spring Harbor Laboratory Press, New York.
17. Vollrath, D. and Davies, R. W. (1987). *Nucleic Acids Res.*, **15**, 7865.
18. Feinberg, A. P. and Vogelstein, B. (1983). *Anal. Biochem.*, **132**, 6.
19. Poustka, A. and Lehrach, H. (1986). *Trends Genet.*, **2**, 174.
20. Pohl, T. M., Zimmer, M., MacDonald, M. E., Smith, B., Bucan, M., Poustka, A., *et al.* (1988). *Nucleic Acids Res.*, **16**, 9185.
21. Michiels, F., Burmeister, M., and Lehrach, H. (1987). *Science*, **236**, 1305.
22. Ramsay, M., Wainwright, B., Farrall, M., Estivill, X., Sutherland, H., Ho, M. F., *et al.* (1990). *Genomics*, **6**, 39.

23. Lichter, P., Tang, C.-J., Call, K., Hermanson, G., Evans, G., Housman, D., and Ward, D. C. (1990). *Science*, **247**, 64.
24. Cox, D. R., Burmeister, M., Price, E. R., Kim, S., and Myers, R. M. (1990). *Science*, **250**, 245.
25. Ho, M. F., Monaco, A. P., Blonden, L. A. J., van Ommen, G. J. B., Affara, N. A., Ferguson-Smith, M. A., and Lehrach, H. (1992). *Am. J. Hum. Genet.*, **50**, 317.
26. Bickmore, W. A. and Sumner, A. T. (1989). *Trends Genet.*, **5**, 144.

Mutation detection and diagnosis using PFGE

J. T. DEN DUNNEN, P. LIANG, G. J. B. VAN OMMEN, and
C. VAN BROECKHOVEN

1. Introduction

Pulse field gel electrophoresis (PFGE) was specifically designed to separate large DNA fragments (1). Consequently, the great innovation of PFGE analysis for mutation identification and diagnosis was its range of detection and its ability to scan for genetic rearrangements at large distances from a given locus (probe). For a specific disease, when closely linked genetic markers become available, PFGE analysis is used to determine the physical map of the region involved. Simultaneous analysis of patient material will quickly reveal if large chromosomal aberrations play a role in the development of the disease. In the case of Duchenne muscular dystrophy (DMD) this has led to the detection of large deletions and duplications as the disease-causing mutation in over 50% of the patients at a time when the gene involved had not yet been identified (2). Similarly, PFGE allowed the detection of large chromosomal duplications in a majority of patients with Charcot–Marie–Tooth disease type 1a (CMT1a), with probes that had been shown to be closely linked by genetic mapping (3, 4). Again, at that time, the responsible gene(s) had not yet been identified.

A second major advantage of PFGE analysis is its ability to detect 'junction fragments', i.e. fragments with an altered size, derived from the fragment(s) carrying the mutation. For instance in duplications, these junction fragments give superior diagnostic resolution compared to dosage comparisons indicating increased signal intensities. Detection of 'junction fragments' with conventional agarose gel electrophoresis (AGE)/Southern blotting is usually rare; conventional analysis of *Hind*III digests detected junction fragments in less than 5% of the DMD patients analysed, while PFGE analysis of *Sfi*I digests revealed junction fragments in over 50% of the cases (2, 5).

Since the theoretical and technical aspects of PFGE as a technique are discussed in Chapters 1 and 2 of this book, the focus here will be on the application of PFGE in mutation detection and diagnosis. The protocols used previously will be provided (6, 7) together with ways to perform quality control. The different classes of chromosomal aberrations are discussed in detail with emphasis on the correct identification of the type and physical length involved. Finally, some examples will be shown and the main problems that occur will be addressed, including rare polymorphisms and differential methylation.

2. Preparation and analysis of DNA samples

2.1 Isolation of DNA in agarose

Essentially, two methods have been described to prepare agarose-embedded DNA of sufficient double-stranded length. The method first described by Schwartz and Cantor (1, 8) of embedding cells in agarose blocks (plugs) is described in *Protocol 1*. The plugs are easy to manipulate, no centrifugations are required, no 'void volume' exists in buffer changes, and no increased quantities of enzymes are necessary. Alternatively, the cells can be included in agarose beads (9).

In principle it is feasible to prepare high molecular weight DNA in solution, yielding DNA of 100–400 kb. However, this method is very laborious since extreme care is necessary to avoid DNA shearing (using cut-off pipette tips, gentle mixing, overnight dissolving of samples, etc.). When enzymes are used which cut very infrequently, like *Not*I, where most of the digested DNA remains above 500 kb, there is no alternative to using agarose-embedded DNA.

Protocol 1. Preparation of agarose blocks containing human DNA

Equipment and reagents

- Blood lysis buffer: 155 mM NH_4Cl, 10 mM $KHCO_3$, 1 mM EDTA, pH 7.4
- Phosphate-buffered saline (PBS): 144 mM NaCl, 10 mM KH_2PO_4 (pH 7.8 with K_2HPO_4)
- SE buffer: 75 mM NaCl, 25 mM Na-EDTA, pH 7.4
- SarE buffer: 1% (w/v) *N*-lauroylsarcosine, 0.5 M EDTA, pH 9.5
- Proteinase K (Boehringer Mannheim) or pronase
- TE: 10 mM Tris–HCl, 1 mM EDTA, pH 7.4

- Low melting temperature (LMT) agarose (InCert-agarose, FMC)
- Centrifuge
- Perspex mould with slots 10 × 6 × 1.5 mm
- Pasteur pipette bulb
- Phenylmethylsulfonylfluoride (PMSF, Sigma) dissolved at 40 mg/ml in isopropanol. (**Caution**: PMSF is toxic and should be handled with gloves.)
- 50 mM EDTA, pH 8.0

Method

1. DNA is usually isolated from white blood cells but other sources can be used and are listed below.

   (a) **Blood**: mix 10 ml heparinized blood with 30 ml blood lysis buffer, leave for 15 min on ice, and centrifuge for 15 min at 1800 *g*.

Resuspend the pellet in 10 ml blood lysis buffer, leave for 15 min on ice, and centrifuge for 15 min at 1800 *g*. On average, 10 ml blood yields enough leucocytes to make about 30 plugs. Store heparinized blood at −70°C before performing DNA isolation.

(b) **Cultured cells**: collect cells after trypsinization using standard procedures.

(c) **Sperm**: follow standard procedure but add 10 mM dithiothreitol (DTT) in the Proteinase-K step.

(d) **Fresh tissue**: dissect the tissue, cut it into small cubes, and homogenize it in PBS to a single-cell suspension, using a glass homogenizer with a tight-fitting pestle. Wash cells once in PBS and resuspend in SE buffer.

(e) **Frozen tissue**: quick-freeze dissected tissues in liquid nitrogen and store at −70°C. Grind 1 g frozen tissue to a fine powder with a pre-cooled pestle and mortar. Complete homogenization to a single-cell suspension in PBS using a glass homogenizer. Wash cells once and resuspend in about 5 ml SE buffer.

2. Wash the cells once in SE buffer and resuspend in SE buffer at 15×10^6 cells/ml, at room temperature.

3. Melt 1% (w/v) LMT agarose in SE buffer, cool to about 50°C, and mix in a 1:1 ratio with the cell suspension.

4. Immediately dispense the mixture (100 μl volume) into slots made through a Perspex mould covered on one side with tape.

5. Put the mould on ice for 5–10 min.

6. Remove the tape and blow the solidified blocks gently out of the slots using a Pasteur pipette bulb.

7. Collect the blocks in about five volumes of SarE buffer.

8. Add Proteinase K to 0.5 mg/ml and incubate for 48 h at 50°C.

9. Incubate the plugs twice for 30 min at 50°C in TE.

10. Rinse the blocks several times with distilled water and wash at least three times for 2 h and once overnight in 10–20 volumes of TE, by gentle rotation.

11. Store the blocks at 4°C in 0.5 M EDTA (pH 8.0).

0.1 mM PMSF can be added to step 9 to block protease activity. Proteinase K can be replaced with pronase (1 h preincubated) at the same concentration, followed by overnight incubation at room temperature, without noticeable effect on the results. Storage of plugs can also be performed in 50 mM EDTA.

Before the isolated DNA can be used check its quality. A smear throughout the lanes after electrophoresis indicates poor cell lysis and DNA degradation. In 'emergency cases' (e.g. rare, valuable samples) remove degraded DNA from a plug by a short pre-electrophoresis before further handling of the sample.

2.2 Restriction digestion of agarose-embedded DNA

Digestion of agarose-embedded DNA is described in *Protocol 2* and requires conditions similar to soluble digestions. However, PFGE analysis requires the use of specific, infrequently cutting restriction endonucleases, modifications of digestion protocols, altered techniques to load the DNA samples, and modification of the techniques to blot and hybridize the DNA. The most frequently used rare-cutting enzymes are *Bss*HII, *Eag*I, *Mlu*I, *Nar*I, *Not*I, *Nru*I, *Sac*II, *Sal*I, and *Sfi*I. Except for *Sfi*I, their recognition sequences all contain one or more CpG dinucleotides and the enzymes are thus sensitive to methylation. Digestions of DNA in plugs are essentially performed according to the manufacturer's recommendations, generally at a 2–3-fold enzyme excess. Per lane on a gel, one half or one third of a plug is required (5–7.5 $\times$ 10^5 cells/lane).

Protocol 2. Restriction digestion of DNA in plugs

Equipment and reagents

- Sterile distilled water
- TE (see *Protocol 1*)
- Scalpel
- Restriction enzymes and buffers according to manufacturer
- Spermidine-trihydrochloride (Sigma, 100 mM stock solution)
- Dithiothreitol (DTT; Sigma, 50 mM stock solution)
- Bovine serum albumin (BSA; Boehringer Mannheim, MB grade, 5 mg/ml stock solution)
- 5 mM EDTA pH 8.0

Method

1. Cut the plugs (approximate volume 100 μl) in half, before use, with a scalpel.

2. Rinse the plugs, which were stored in EDTA, once with sterile water and wash extensively in TE (i.e. at least three times for 2 h in 10–20 volumes under gentle rotation).

3. Equilibrate each 50 μl plug for 2 h at room temperature or overnight at 4°C with 1 ml of the appropriate restriction enzyme digestion buffer.

4. Replace the equilibration buffer by 50 μl fresh restriction enzyme digestion buffer, to which 4 mM spermidine, 2 mM DTT, and 0.2 mg/ml BSA have been added.

---

5. Carry out digestions for 6 h to overnight at the specified temperature, using 10–20 units of enzyme per 50 µl plug. Add the enzyme in two equal portions, at the beginning and half way through the digestion time.

6. After digestion, layer the plugs directly or store in 5 mM EDTA at 4°C.

For double digestions; repeat steps 1 to 4 for the second enzyme. For the second digestion step 1 may be preceded by a Proteinase K treatment, but this should not be necessary. For digestions with frequently cutting restriction endonucleases (like *EcoR*I, *Hind*III) modify the protocol after step 3 to carefully remove all equilibration buffer and incubate for 10 min at 65°C to melt the plug. Incubate for 15 min at 37°C, directly add DTT, spermidine, BSA, and restriction enzyme, and incubate at the desired temperature. Layer the melted plug directly on to the gel.

Before a first digestion, a control incubation should be performed without the addition of enzyme. After the incubation, DNA degradation should be negligible in the size range under study (normally below 2 Mb). Remaining DNA degrading activities should be eliminated by a second Proteinase K treatment.

2.3 PFGE analysis

The PFGE system we use has either the Pulsaphor 2015 (LKB) (7) or the Gene-tic power supply (6) to drive the electrophoresis. Temperature variations have a pronounced effect on the separation patterns obtained with PFGE. Since the high voltage used during electrophoresis generates a considerable amount of heat, temperature is controlled at either 12°C (7) or 18°C (6) by a thermostatic circulator.

2.3.1 Gel preparation

The preparation of gels for PFGE is described in *Protocol 3*.

Protocol 3. Gel preparation

Equipment and reagents

- 0.5 × TBE buffer: 45 mM Tris-borate, 1 mM EDTA, pH 8.3
- 1% (w/v) agarose (SeaKem GTG, FMC) in 0.5 × TBE buffer
- Pasteur pipette and scalpel blade
- PFGE apparatus

Method

1. Prepare a 1% (w/v) agarose gel in 0.5 × TBE buffer.

2. Thoroughly remove the supernatant of the plugs containing digested DNA.

Protocol 3. *Continued*

3. Transfer the plugs to the gel slots prefilled with 0.5 × TBE buffer, using a Pasteur pipette bent into a hairpin and a scalpel blade.

4. Similarly, transfer the plugs containing marker DNAs to the gel slots.

5. Equilibrate the gel with 0.5 × TBE buffer in the gel tray for 30–60 min.

6. Start the electrophoresis: 35 h at 150 V and 12 °C with a switch time of 60 sec to separate fragments of 100–1000 kb, or 20 sec to separate from 100–300 kb.

Fix the plugs in the gel slots with agarose or alternatively, melt the samples for 5 min at 65 °C and gently pipette it into the gel slots, using a cut-off pipette tip. Since some shearing of the DNA will occur, the latter procedure can not be recommended when very large fragments (over 1 Mb) have to be detected. Moreover, occasional degradation of the DNA might occur during the melting step.

Preferred DNA size markers are concatemers of bacteriophage λ (50–600 kb) or the chromosomes of *Saccharomyces pombe* (200–1500 kb), *Kluveromyces lactis* (900–3000 kb), and *Schizosaccharomyces pombe* (3000–6000 kb). These markers can be easily prepared (6) or obtained directly from several commercial suppliers (see Chapter 2).

2.3.2 Staining and blotting of PFGE gels

Blotting of large DNA molecules separated on PFGE gels requires efficient DNA transfer. Individual labs prefer either acid depurination or UV irradiation to reduce DNA size (6, 7). The staining and Southern blotting of PFGE gels is described in *Protocol 4*.

Protocol 4. Staining and blotting of PFGE gels

Equipment and reagents

- Denaturation buffer: 0.15 M NaCl, 0.5 M NaOH
- Neutralization buffer: 0.5 M Tris–HCl, 0.15 M NaCl, 1 mM EDTA, pH 7.2
- 1 × SSC: 150 mM NaCl, 15 mM Na$_3$-citrate
- 2 μg/ml ethidium bromide
- 0.25 M HCl
- Distilled water
- Nylon membrane (Hybond N+, Amersham)
- 0.4 M NaOH
- 0.2 M Tris–HCl pH 7.5
- Stratagene Stratalinker or UV-light trans-illuminator

Method

1. Stain the gel for 0.5–1 h in 2 μg/ml ethidium bromide.

2. Photograph the gel immediately or destain by washing extensively in water to improve the contrast.

3. Place the gel 5 min in 0.25 M HCl, rinse with distilled water, and denature 30 min in denaturation buffer. Rinse with water and wash twice for 15 min in neutralization buffer.

4. Blot the gel, upside down, overnight (at least) in 10 × SSC to Hybond N+ membrane.

5. Change the paper towels in the morning, 1 h before dismounting the blotting stack.

6. Immobilize the DNA by soaking the membrane (DNA-side up) for 10 min on a puddle of 0.4 M NaOH. Wash 15 min in 2 × SSC with 0.2 M Tris–HCl (pH 7.5). Dry the blot for 1 h at 65°C.

For UV irradiation change step 3 to: irradiate the gel (upside down) at 180 000 $\mu J/cm^2$ in the Stratalinker (Stratagene) and wash twice for 15 min in 0.4 M NaOH (6). Alternatively use a UV transilluminator either for 60–90 sec with 254 nm light, or 5–10 min with 302 nm light. It is essential, however, to test each UV transilluminator to define the optimal illumination time.

2.3.3 Hybridization of PFGE blots

Perform the hybridization of PFGE blots using standard hybridization protocols, with one of the methods described in *Protocol 5*. In our experience, cDNA or small genomic probes give excellent hybridization results. Competitive DNA hybridization using whole cosmids also gives good results but requires initial testing to find the optimal conditions for efficient competition of the repetitive DNA sequences in the probe (10). Theoretically, cosmids spanning a restriction site of the enzyme used to digest the DNA are preferable probes since more information can be obtained from one hybridization (see below).

Protocol 5. Hybridization

Equipment and reagents

- Hybridization buffer: 0.125 M Na_2HPO_4 (pH 7.2 with H_3PO_4), 0.25 M NaCl, 1.0 mM EDTA, 7% (w/v) SDS (Sigma, 44244), 10% (w/v) PEG 6000 (BDH Chemicals Ltd)
- Radioactive labelling kit (Multiprime kit, Amersham)
- Sephadex G50 column in a Pasteur pipette
- Wash solutions requiring SSC (*Protocol 4*) and SDS
- Sheared human placental DNA (Sigma)

Method

1. Label 10 ng DNA (Multiprime kit, Amersham). Purify over a Sephadex G50 column in a Pasteur pipette. Boil for 5 min and chill on ice.

2. Prehybridize the membranes in hybridization buffer for at least 10 min at 65°C in a water bath.

Protocol 5. *Continued*

3. Add the denatured label to the prehybridized membranes, mix thoroughly, and hybridize overnight at 65°C.

4. Wash the membranes from 2.0 × SSC / 0.1% (w/v) SDS (twice for 15 min), 1.0 × SSC/0.1% (w/v) SDS (twice for 15 min) and, when necessary, down to 0.3 × SSC/0.1% (w/v) SDS (once for 15 min) at 65°C.

5. Autoradiography takes 4 h–3 days at −70°C using an intensifying screen.

For competitive hybridization (10): step 1: transfer half of the G50 eluate (c. 200 μl) to an Eppendorf tube. Add 240 μl competitor DNA (500 ng/μl human placenta DNA sonicated to 100–1000 bp). Boil for 5 min and chill on ice. Add to a capped 10 ml plastic or polypropylene tube with 1.5 ml hybridization buffer, preheated to 65°C). Mix thoroughly and incubate 90 min at 65°C in a water bath (**NB**: time is critical!).

2.3.4 Rehybridization of PFGE blots

Blots hybridized with a specific probe can be analysed with other probes after removal of the old hybridization signal ('stripping') as outlined in *Protocol 6*. The method should be as gentle as possible for the DNA to remain fixed to the membrane.

Protocol 6. Stripping PFGE blots

Equipment and reagents

- Strip-mix A: 20 mM Na_2HPO4 (pH 7.2 with H_3PO_4), 0.5 mM EDTA, 50% (w/v) formamide, 0.5% (w/v) SDS, 0.5 × SSC
- Strip-mix B: 0.01 M NaOH

Method

1. Wash 30 min at 65°C in strip-mix A under gentle rotation.

2. Wash 15 min in 0.1 × SSC/0.1% (w/v) SDS and air-dry the filter.

Alternatively, filters can be stripped in an alkaline solution (strip-mix B): wash 15 min in 0.01 M NaOH, followed by a 15 min wash in 2 × SSC/0.2M Tris–HCl (pH 7.5). Stripped filters should be checked by overnight autoradiography to detect old signal.

3. Detection of chromosomal rearrangements

The probes that are used most frequently in PFGE analysis are located somewhere within the fragment detected. Such probes have the intrinsic danger of missing some rearrangements. In deletion mutations, probes located in the deletion will give no signal and thus fail to detect the mutation. In the example shown (*Figure 1*, left deletion), only probes located directly downstream of site C or directly upstream of site D will reveal the deletion mutation. To safeguard against mistakes, each fragment should be analysed

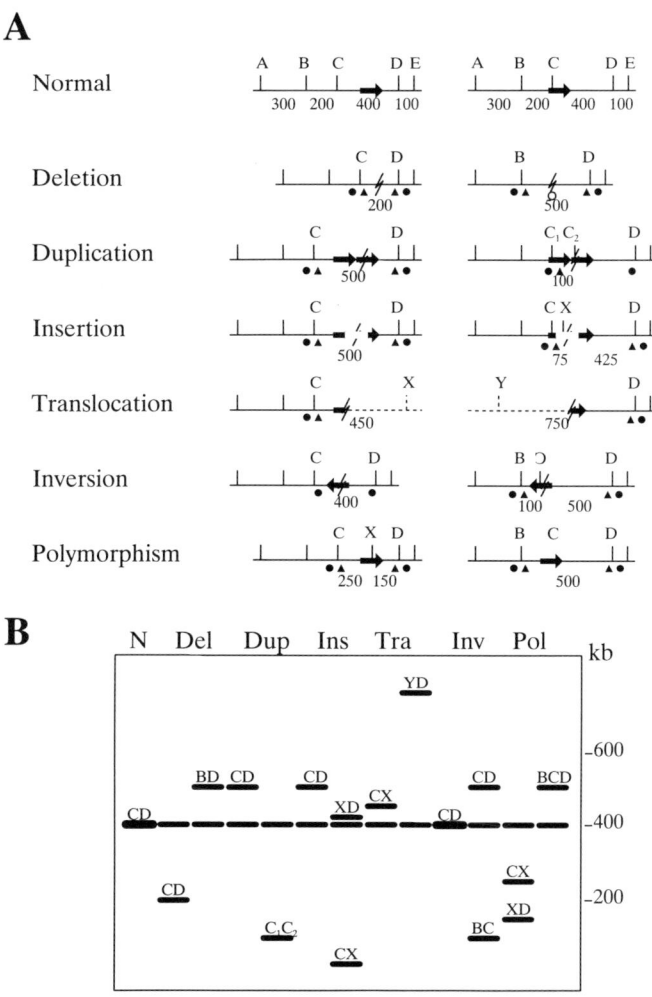

Figure 1. Chromosomal aberrations and their effect on a PFGE analysis. (a) Physical map with imaginary mutations either with (right) or without (left) the involvement of a restriction site (affected region indicated by an arrow). Shown are a normal situation, deletion, duplication, insertion, translocation, inversion, and polymorphism. Individual restriction sites are marked with letters (A–E). Filled triangles indicate probes detecting altered fragments while filled circles mark probes giving normal hybridization patterns. (b) Schematic drawing of a PFGE analysis of the mutations given in (a). Sizes are indicated in kb, letters mark the hybridizing fragments.

with two probes, one flanking the 5' and one flanking the 3' end. A proper selection of the hybridization probes used will considerably speed up the confirmation of the type of rearrangement detected. Probes comprising a restriction site are preferred, since they detect two flanking fragments in one hybridization. In the examples drawn (*Figure 1*), probes spanning sites C and D will in most cases be sufficient to complete the analysis. Although such probes, usually cosmids, have to be used in combination with competitive DNA hybridization protocols (10), this should not cause major problems. Other probes spanning a restriction site, e.g. cDNA probes of multi-exonic genes, are also very efficient.

Chromosomal rearrangements detectable with PFGE can be divided into several types (*Figure 1*); deletions (missing sequences), duplications (doubled sequences), insertions (extra sequences), translocations (sequence exchange between chromosomes), and inversions (sequences turned around). Every rearrangement will have a characteristic effect on the hybridization pattern obtained and the emerging picture is affected significantly by the presence or absence of a restriction site within the mutation. A PFGE analysis can not be completed unless the existence of phenotypically silent mutations, amongst which are rare polymorphisms, has been ruled out.

3.1 Deletions

Deletions within a restriction fragment result in the appearance of one 'new' band. This new band has a decreased size and can be detected with probes flanking both ends (*Figure 1* left deletion, probes C and D). Deletions encompassing a restriction site also create one abnormal fragment, now detectable by probes that normally hybridize with different fragments (*Figure 1* right deletion, probes B and D). Depending on the size of the deletion, this junction fragment is always smaller than the total size of the parental fragments. Probes from within the deletion will not detect the altered fragment.

3.2 Duplications

Duplications within a fragment result in one altered fragment, detectable by the appearance of a band with an increased size. The new band can be detected with probes flanking both ends of the fragment (*Figure 1* left duplication, probes C and D). Duplications including a restriction site create a more complex picture and are more difficult to detect. All fragments within the duplication do not change in size but give a higher signal intensity which is hard to determine with certainty. Only one abnormal fragment will arise which can be visualized by a probe from within the duplication only (*Figure 1* right duplication, probe C).

3.3 Insertions

Insertions within a fragment result in increased fragment size, detectable as one new fragment with a decreased gel mobility. The new band can be detected with probes flanking both ends (*Figure 1* left insertion, probes C and D). Such insertions can be discriminated from duplications only by the fact that no probe can be found which gives a higher signal intensity (i.e. is duplicated). Insertions adding a new restriction site will split a fragment, detectable by two probes, into two separate fragments. It is characteristic that the total size of the two new fragments is always larger than the size of the original fragment.

Insertions are usually caused by the integration of repetitive DNA elements, like Alu and LINE repeats, at new chromosomal locations. Consequently, they are rather small, i.e. 0.3–6 kb, and can not be detected easily with PFGE but only by conventional AGE analysis in combination with Southern blotting. Although most insertions occur in intergenic or intronic regions and are phenotypically silent (i.e. create restriction fragment length polymorphisms, RFLPs), disease-causing insertions have also been identified (11, 12).

3.4 Translocations

Translocations are detected by the appearance of two altered fragments since they split an existing fragment. Probes which are localized on one fragment will thus hybridize with two new fragments having an unpredictable size (*Figure 1* translocation, probes C and D). Unequivocal proof of a translocation mutation requires the availability of a probe on the other site of the translocation break point (*Figure 1* translocation, probe X or Y). Since these probes are usually not available, microscopic analysis of metaphase chromosome spreads will in such cases be a much simpler alternative.

Translocations occur frequently in several types of genetic disease. Many types of cancer are caused by specific interchromosomal translocations (13, 14), while in other types of cancer translocations are found only occasionally (15, 16). In Duchenne muscular dystrophy, a recessive X-linked disorder, females manifesting the disease frequently carry X/autosome translocations (17, 18). Finally, translocations occur frequently when different chromosomes contain large stretches of homologous sequences, e.g. X/Y translocations (19).

Usually translocations are detected in cytogenetic analysis of patient cells. When close physical markers are available, PFGE analysis can be used to localize precisely the translocation break points, to determine their physical distribution, and, finally, to clone the gene which is disrupted. Recently, fluorescence *in situ* hybridization (FISH) combined with yeast artificial chromosomes (YACs) probes was developed as an attractive alternative tool to detect translocations (14, 20).

3.5 Inversions

Inversions reverse the orientation of the sequences involved. Inversions within a specific fragment will not change the hybridization pattern obtained

(*Figure 1*, left inversion). Consequently, inversions can only be detected when they cross a restriction site. In such cases, two probes residing on flanking fragments will both detect one new fragment (*Figure 1* right inversion, probes B and D). In contrast to translocations, the sum of the sizes of these two fragments will equal the sum of the original fragments. Definitive proof of inversion mutations requires the availability of probes flanking both ends of the inversion.

Until recently, inversion mutations were only found in a few exceptional cases (21, 22). However, Naylor *et al.* (23) and Lakich *et al.* (24) showed that inversions in the factor VIII gene are responsible for the majority of cases of severe Haemophilia A. Since one break point resides in the 32 kb intron 22 and the other side some 500 kb 3' (telomeric) of the gene, these inversions had not been detected using factor VIII cDNA probes in combination with conventional gel electrophoresis. Once the inversion had been established, a PFGE analysis clearly demonstrated its ability to detect the inversion (23).

4. Examples of PFGE detection of chromosomal rearrangements

Initially, PFGE can be used to scan for large genetic rearrangements in the vicinity of a marker without any knowledge of the gene(s) involved. Ultimately, however, a precise physical map of the gene involved will be the basis for the application of PFGE to detect mutations. The physical map is constructed using a combination of single and double enzyme digests with several rare-cutting restriction endonucleases. The borders of the target gene should be defined as precisely as possible and one or two rare-cutting restriction enzymes should be selected which cover the whole region in one (or a few) clearly detectable fragments. This selection is then combined with the choice of specific electrophoretic conditions which optimize the detection of size differences in the fragment(s) under study.

Most examples presented below will be taken from the analysis of Becker and Duchenne muscular dystrophy (BMD/DMD), an X-linked recessive disorder caused by mutations in the dystrophin gene (25). PFGE analysis of the human dystrophin gene revealed two main features: (i) the gene is extremely large, measuring 2.4 Mb (5), and it can not be covered by one restriction fragment; (ii) most restriction enzymes fail to give clearly detectable fragments for all parts of the gene. *Sfi*I appeared to give the most reproducible results. The *Sfi*I map of the DMD gene contains nine intragenic sites (*Figure 2*). Five clearly discernible restriction fragments are present in the size range of 200–850 kb, i.e. fragments AC, CD, DF, FI, and IJ. Three of these fragments contain partially digestible sites (*Figure 2*), which, although complicating the hybridization patterns obtained, are excellent tools to define the type and borders of the rearrangements detected (see below).

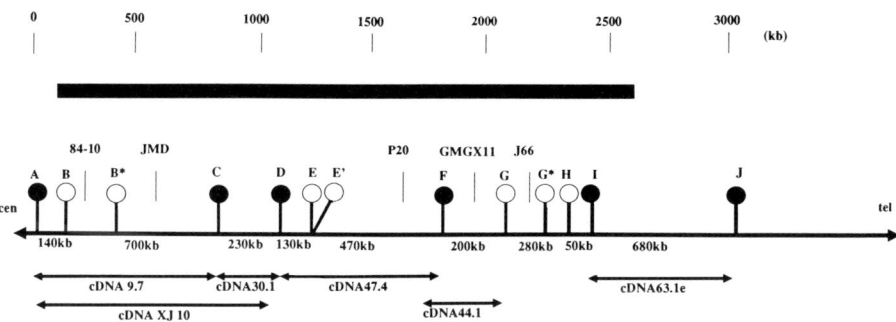

Figure 2. Schematic long-range physical map of the human dystrophin gene (horizontal thick bar) showing all partially (open circles) or fully (closed circles) digestible *Sfi*I sites (6, 7). Location and/or extent of the probes used to detect the fragments are indicated. Individual *Sfi*I sites are marked with letters (A–J); sites marked by an asterix are polymorphic (7, 29). Fragment lengths are indicated in kb.

Fragment EE′ measures less than 10 kb and will not be detected in a PFGE analysis unless special precautions are taken. Five probes are necessary to cover the gene, i.e. XJ10, cDNA47.4, cDNA44.1, J66, and cDNA63.1e. Since DMD is an X-linked disease, the analysis of male patients with intragenic probes will detect sequences from only one chromosome.

4.1 Examples of deletions

In DMD family 124, hybridization of cDNA44.1 to DNA of the patient detects a smaller *Sfi*I-fragment FG (170 kb) next to a normal fragment EF (*Figure 3*). Since fragment GH is not altered (not shown), the patient carries a 30 kb deletion in fragment FG of his DMD gene. Both his mother and her sister have, next to a normal 200 kb fragment FG, the altered FG band as well. This indicates that the DNA of both females contained the same deletion and thus these women are DMD carriers. cDNA Southern blot analysis of *Hind*III-digested patient DNA confirmed the presence of a deletion (i.e. deletion of exon 51).

In the second example, no patient material was available for analysis in DMD family 40. Initial haplotype analysis was not conclusive, since a recombination was detected in the DMD gene. Elevated CK values were measured in both the mother and the two at-risk females, suggesting that they carried a mutation in the DMD gene. PFGE analysis (*Figure 4*) on DNA of the three females showed abnormal hybridization patterns with cDNA44.1 and genomic marker P20. The abnormal pattern constituted an extra band of 550 kb. The extra band can be detected by probes located on both sides of *Sfi*I–F, which should thus be involved in the mutation. Since no differences in signal intensities can be observed we concluded that the pattern arises from a 120 kb deletion around *Sfi*I–F. The deletion could be confirmed by a conventional

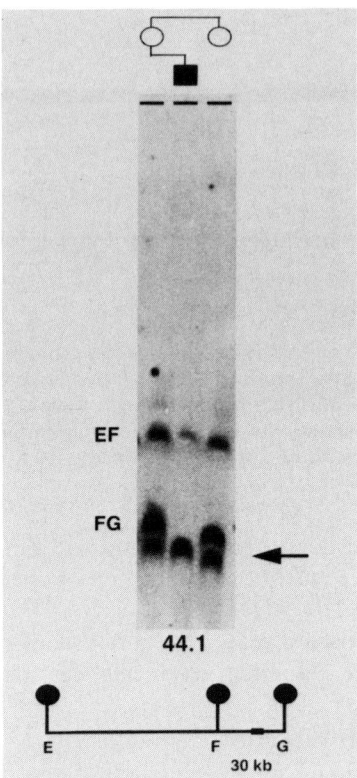

Figure 3. PFGE analysis of DMD family 124. *Top*: PFGE blot hybridized with cDNA44.1 showing an altered *Sfi*I fragment FG (arrow) present in the patient, his mother, and his aunt. *Bottom*: schematic picture illustrating the location and size of the deletion mutation detected.

Southern analysis with cDNA44.1; dosage comparison in the carriers indicated a deletion of exons 47–50.

Large deletions have been detected in many diseases, although in highly variable percentages. For DMD, large deletions are the most frequent type of mutation observed; they constitute about 55% of all mutations. Another very recent example in which PFGE analysis was used to prove the involvement of large genomic deletions was Fascioscapulohumeral dystrophy (FSHD) (26, 27). Although the large FSHD deletions can be used as a diagnostic tool, the gene(s) involved has not yet been identified.

4.2 The power of junction fragments; carrier detection

The pedigree of family 56, shown in *Figure 5*, contains three affected boys who all died of DMD before the genetic analysis was started. PFGE analysis

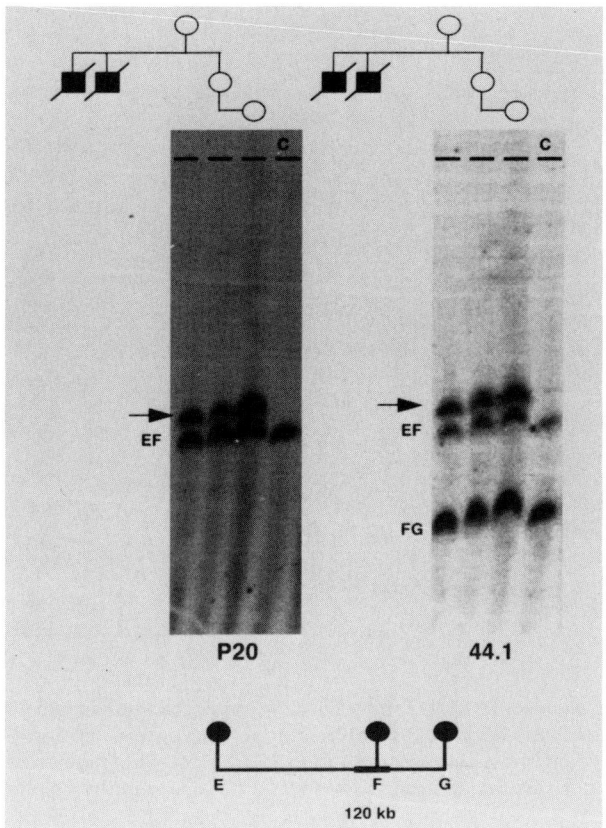

Figure 4. PFGE analysis of DMA family 40. *Top*: PFGE blot showing the extra band (arrow) detected by probes P20 (left) and cDNA44.1 (right). *Bottom*: schematic picture illustrating the location and size of the deletion mutation detected.

of the mother (individual 1) and the sister of the patients (individual 2) detected abnormal hybridization patterns with cDNA44.1 and genomic marker J66. Dual signals were obtained for the *Sfi*I fragment FG and FH, suggesting that a 50 kb deletion had occurred between *Sfi*I sites F and G (*Figure 5*). Hybridization with genomic marker GMGX11 on the same PFGE blot (not shown) shows a normal hybridization pattern. This indicates that GMGX11 is located in the deleted region and demonstrates the danger of the use of such probes, i.e. small genomic probes not directly flanking a restriction site, for this type of analysis; it failed to detect the deletion.

Hybridization with cDNA44.1 on a conventional Southern blot did not reveal any dosage differences of exon-containing fragments. Consequently, the exons involved in the deletion could not be identified. Since both mother and daughter (*Figure 5*) showed the same abnormal hybridization pattern we

59

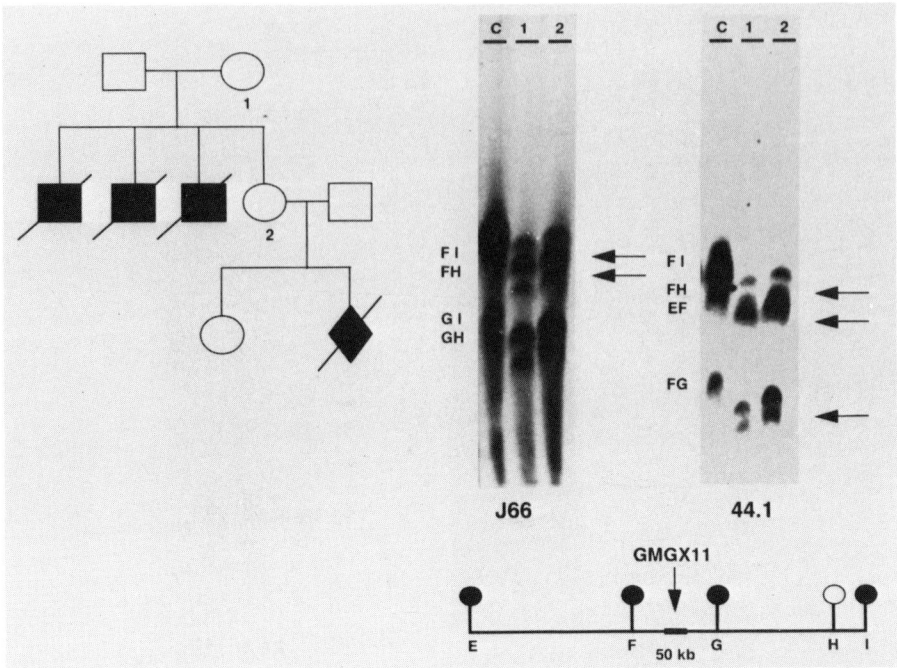

Figure 5. PFGE analysis of DMD family 56. *Left*: pedigree. Individuals 1 and 2, analysed by PFGE, are indicated. *Right*: PFGE blot hybridized with probes J66 and 44.1 respectively. Both females tested show dual hybridization signals for *Sfi*I fragments FG, FH, and FI. *Bottom*: schematic picture illustrating the location and size of the deletion detected.

concluded that they were carriers of the deletion. Two prenatal diagnoses were performed in the family, of which the second was on a male fetus. Southern blot hybridization revealed a deletion of exons 49 and 50 and probe GMGX11 in the fetal DNA, thus confirming our earlier PFGE results.

4.3 Examples of duplications

PFGE analysis of DMD family 43 showed that fragments AC and BC, detected by cDNA probe XJ10, were increased in size (*Figure 6*). Since no other additional fragments were detected and since fragments AB and CD were not affected, we concluded that a 200 kb duplication is responsible for the hybridization pattern observed. The power of PFGE analysis in comparison with conventional AGE analysis is demonstrated nicely with the results presented (*Figure 6*). Although the high quality of the Southern blot does allow a confirmation of the presence of a duplication (compare individuals 5 and 6), the exact exon borders can not be determined. Moreover, only the coincidental presence of a duplication junction fragment allows a clear identification of individuals 2, 4, 6, and 7 as carriers of the mutation. However, the

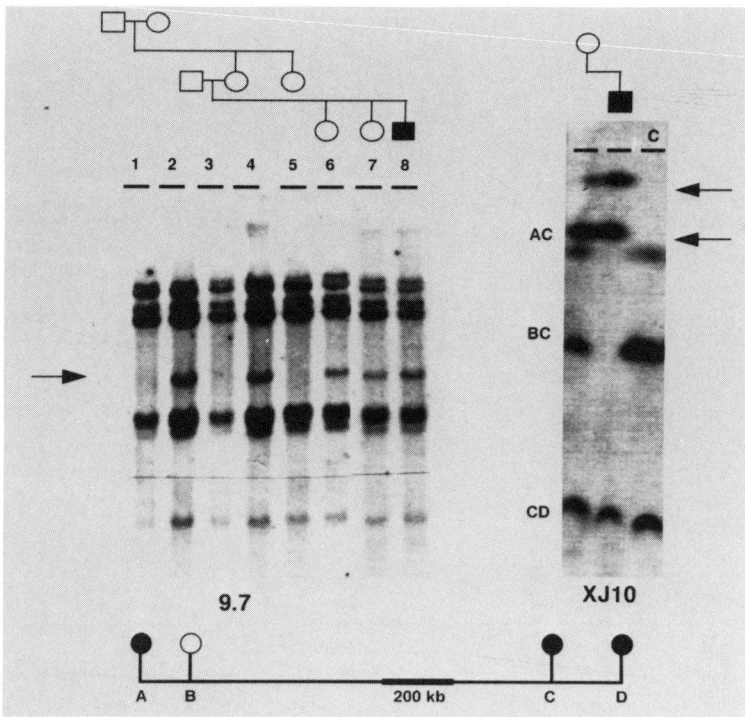

Figure 6. Comparison of conventional AGE analysis (left panel) and PFGE analysis (right panel) of DMD family 43. *Left*: Southern blot hybridized with probe cDNA9.7. The duplication junction fragment detected in *Pst*I-digested DNA of the patient and the carriers is indicated (arrow). *Right*: PFGE-blot hybridized with probe cDNAXJ10 showing duplication junction fragments in both the patient and his mother. *Bottom*: schematic picture illustrating the location and size of the duplication detected.

presence of such duplication-junctions fragments is rare (for DMD less than 5% of the cases (7)). The PFGE analysis (*Figure 6*) gives a much clearer picture. Dosage comparisons in conventional AGE analysis usually allow detection of deletions, resulting in 1:2 signal ratios, but detection of duplications, giving signal ratios of 3:2, is hardly possible. For duplication analysis, PFGE should be the preferred method.

Duplication mutations are not specific for DMD, they can also be detected in other diseases. Recently, a duplication of a 1.5 Mb region of chromosome 17p11.2 was detected as the major cause of Charcot–Marie–Tooth disease type 1a (CMT 1a) (3, 4, 28). The CMT1a duplication can be detected by Southern blot hybridization of *Msp*I-digested DNA with probe pVAW409R3a (D17S122). Since the latter probe detects a *Msp*I RFLP the duplication can be detected in heterozygous individuals as a density difference between the polymorphic alleles (3). The duplication can also be detected by the presence

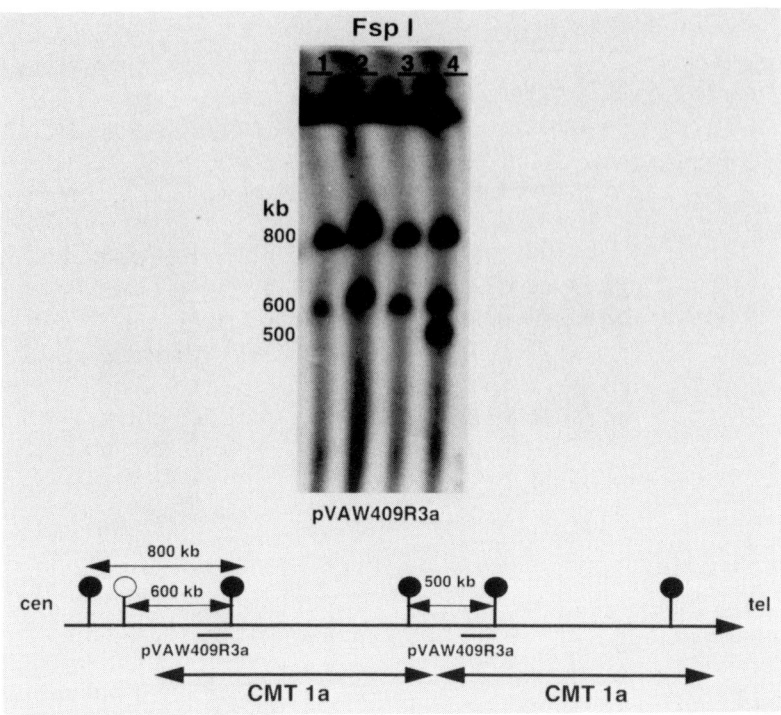

Figure 7. PFGE detection of DMT1a duplications. *Bottom*: *Fsp*I long-range physical map of the 1.5 Mb CMT1a duplication with individual sites indicated as in *Figure 2*. The localization of the hybridization probe pVAW409R3a, the hybridizing fragments observed, and the CMT1a duplicated region are indicated. *Top*: Southern hybridization of probe pVAW409R3a showing the 500 kb *Fsp*I duplication junction fragment detectable in CMT1a patients only.

of three alleles of a CA-repeat polymorphism associated with locus D17S122 and located within the duplication (4). However, since both of these detection methods rely on the heterozygosity of the polymorphic marker used, the duplication will remain undetected in a number of cases. This problem can be overcome by direct detection of the CMT1a duplication by PFGE (*Figure 7*). Hybridization of *Sac*II-, *Asc*I-, or *Fsp*I-digested DNA of CMT1a patients with probe pVAW409R3a will demonstrate an extra band of 500 kb (28), the duplication junction fragment. We routinely use *Fsp*I-digested DNA separated in the 100 to 1000 kb range (*Figure 7*). In the example shown, only individual 4, who is a CMT patient, shows the 500 kb junction fragment and therefore carries the 1.5 Mb CMT1a duplication. Individuals 1 to 3 were either at-risk family members or patients with a CMT-related phenotype; they do not carry the CMT1a duplication.

5. Potential pitfalls

5.1 Polymorphisms

Although the detection of an altered fragment is the ultimate aim of a PFGE analysis, detection of an altered fragment only is never sufficient to conclude that a genetic rearrangement has occurred. First, as explained, analysis of the flanking fragments is essential to establish the type of rearrangement involved. Second, there is always the danger of rare restriction fragment length polymorphisms (RFLPs) obscuring the analysis. *Figure 1* shows the two most frequent types: point mutations resulting in the absence of a site or the occurrence of a new site. The absence of a site is easily detectable since probes that normally detect different fragments now hybridize to one shared fragment (polymorphism *Figure 1*, probes, B, C, and D). The size of this new fragment characteristically equals the sum of the 'missing' individual fragments. The occurrence of an extra site gives the reverse situation. Probes residing on one fragment now detect two different fragments of which the added size equals that of the original fragment (*Figure 1* right polymorphism, probes C and D).

RFLPs originating from large insertions, although their occurrence is probably very rare (see above), may cause more problems. Depending on whether the insertion introduces a new restriction site or not, such RFLPs can easily be mistaken for translocations or duplications respectively (see *Figure 1* left deletion or *Figure 1* translocation).

There are several ways in which RFLPs can be discriminated from disease-causing chromosomal aberrations. First, RFLPs are detectable in unrelated individuals and thus also in unaffected persons. Second, RFLPs do not segregate with the disease phenotype. Third, the detection of several RFLP types is specific for one enzyme; they can not be detected when a second enzyme is used.

5.2 Tissue-specific methylation patterns

Rare-cutting restriction enzyme digestions often give multiple hybridizing DNA fragments. Since overdigestion has no effect, this is not due to incomplete digestion but due to partial resistance of the sites to cleavage. The most likely cause of this resistance is partial methylation of CpG dinucleotide(s) present in the recognition sequence of most rare-cutting restriction enzymes. The use of restriction enzymes which do not contain CpG dinucleotides in their recognition sequence greatly reduces this problem. Consistent with its site lacking a CpG dinucleotide (i.e. 5'-GGCCNNNNNGGCC-3'), *Sfi*I mostly gives complete digestion patterns. Still, differences can be observed in the digestibility of individual *Sfi*I sites. No data exist on the nature of these differences. One possible explanation is that *Sfi*I is sensitive to methylation of either C_4 or C_{13} in the recognition sequence when this C is followed by a

G. Provided, however, that the pattern becomes not too complex, partial digestion gives additional information 'free of charge': it extends the detection range and permits the scanning of several adjacent fragments.

Methylation patterns are often tissue specific and care should be taken when hybridization patterns obtained from blood-derived patient DNA are compared with existing long-range physical maps, usually derived from DNA isolated from cultured cells. For the DMD gene, several differences can be observed when the hybridization patterns obtained with DNA of fresh lymphocytes are compared with those of EBV-transformed lymphoblasts (6, 7). In lymphocyte DNA, several fragments are either missing or hybridize weakly while extra fragments appear that are not detected in lymphoblast DNA (*Figure 2*). In lymphocyte DNA, *Sfi*I fragments AC, DF, FH, and FI are missing, indicating that the sites B, E, G, and H changed from partially digested sites into completely digested sites. In our experience, sites G and H sometimes cut partially leading to weak hybridization signals for bands FH and FI. Another striking difference is the presence of two additional *Sfi*I sites which can be observed in leucocyte DNA only. One of them, designated site B*, was described recently in both genomic (29) and YAC DNA (30) and is located 260 kb downstream of *Sfi*I-B. The presence of this site results in the detection of a 260 kb fragment BB* with pERT84.10 and a 440 kb fragment B*C with probe JMD (*Figure 2*). The second site, G*, has not been reported before and is located 160 kb downstream of *Sfi*I-G resulting in a 360 kb fragment FG* recognized by probe GMGX11 (7). We decided that these extra sites were not caused by a mutation since they were also observed in normal individuals and segregated within the families. In a total of 43 families investigated by PFGE, site B* was observed in 12 families (28%) and site G* in 6 families (14%).

One exceptional case has been described where methylation differences could be related directly to a disease phenotype. In fragile-X syndrome, expansion of a CGG trinucleotide in the FMR1 gene could be demonstrated as the genetic cause of the disease (31). The expansion, which is usually too small to be detectable by PFGE, causes the inactivation of the gene involved. As a consequence, a *Bss*HII site in the 5′ region of the gene is methylated. This methylation prevents digestion of the site and results in a shift of the *Bss*HII fragment detected from 600 kb to over 1.0 Mb (32), a shift easily detectable by PFGE!

5.3 Miscellaneous

As with any technique, pulsed field gel electrophoresis has its limitations. PFGE is very sensitive to differences in the concentration of the DNA loaded; higher concentrations result in slower migration (e.g. compare *Figure 5* lanes c and 1). Care should thus be taken when lane-to-lane comparisons are performed or when fragment sizes are estimated.

A second limitation of PFGE is its capability to detect small alterations.

Under standard conditions, for lane-to-lane comparisons, alterations of 20–30 kb should be detectable. In heterozygous carriers, detection of even smaller size differences is possible since dual signals are obtained; one from the normal and one from the mutated chromosome. In specific cases, the detection range can be increased when separation conditions are used which optimize the resolution of fragments in a specific size range. In the examples of the DMD gene presented above, small deletions in *Sfi*I fragment CD (*Figure 2*) can be detected when smaller switch times are used during electrophoresis (see Section 2.3.1) (7).

Pulsed field gel electrophoresis is a valuable and powerful tool to scan large genomic regions for chromosomal rearrangements. PFGE is particularly useful to study the presence of rearrangements when only close-flanking markers are available (2, 3), to study specific types of mutations (e.g. duplications), and to detect mutations in large genes (spanning 50 kb or more). Still, a major barrier seems to exist for the general application of PFGE in mutation detection. Devotees of PFGE were probably astonished by the recent discovery that most of the severe cases of haemophilia A are caused by a large 600 kb inversion (23, 24). These inversions disrupt the factor VIII gene, a gene cloned in 1984 and measuring 189 kb — this, in particular, combined with the frequent occurrence of deletion mutations, should have provided an excellent qualification for the use of PFGE for mutation analysis in which case, as demonstrated by Naylor *et al.* (23), the inversion would have been detected years earlier. Why it was not is unclear. Probably, PFGE, and PFGE analysis of human genomic DNA in particular, is considered by many as a complicated, specialist technique rather than a generally useful tool which should be included amongst the standard outfit of a modern laboratory. Hopefully, the results described in this chapter may help to persuade those who have not yet implemented PFGE for mutation detection and diagnosis.

Acknowledgements

This work was supported in part by grants from the Muscular Dystrophy Group of Great Britain and Northern Ireland, the Muscular Dystrophy Association of America, the Dutch Prevention Fund, and the Belgian National Fund for Scientific Research (NSFR). CVB is a research associate of the NSFR.

References

1. Schwartz, D. C. and Cantor, C. R. (1984). *Cell*, **37**, 67.
2. Den Dunnen, J. T., Bakker, E., Klein-Breteler, E. G., Pearson, P. L., and Van Ommen, G. J. B. (1987). *Nature*, **329**, 640.
3. Raeymaekers, P., Timmerman, V., Nelis, E., De Jonghe, P., Hoogendijk, J. E., Baas, F., *et al.* (1991). *Neuromusc. Disord.*, **1**, 93.

4. Lupski, J. R., Montes De Oca-Luna, R., Slaugenhaupt, S., Pentao, L., Guzzetta, V., Trask, B., *et al.* (1991). *Cell*, **66**, 219.
5. Den Dunnen, J. T., Grootscholten, P. M., Bakker, E., Blonden, L. A. J., Ginjaar, H. B., Wapenaar, M. C., *et al.* (1989). *Am. J. Hum. Genet.*, **45**, 835.
6. Den Dunnen, J. T., Grootscholten, P. M., and Van Ommen, G. J. B. (1993). In *Human genetic disease analysis: a practical approach* (ed. K. E. Davies), pp. 35–58. Oxford University Press, Oxford.
7. Liang, P. (ed.) (1993). In *Analysis of dystrophin gene mutations by pulsed field gel electrophoresis*. PhD Thesis, University of Antwerp, Antwerp.
8. Smith, C. L. and Cantor, C. R. (1987). In *Methods in enzymology* (ed. R. Wu), Vol. 155, pp. 449–67. Academic Press, New York.
9. Cook, P. R. (1984). *EMBO J*, **3**, 1837.
10. Blonden, L. A. J., Den Dunnen, J. T., Van Paassen, H. M. B., Wapenaar, M. C., Grootscholten, P. M., Ginjaar, H. B., *et al.* (1989). *Nucleic Acids Res.*, **17**, 5611.
11. Dombroski, B. A., Mathias, S. L., Nanthakumar, E., Scott, A. F., and Kazazian, H. H., Jr (1991). *Science*, **254**, 1805.
12. Mitchell, G. A., Labuda, D., Fontaine, G., Saudubray, J. M., Bonnefont, J. P., Lyonnet, S., *et al.* (1991). *Proc. Natl. Acad. Sci. USA*, **88**, 815.
13. Rowley, J. D. (1982). *Science*, **216**, 749.
14. Liu, P., Tarle, S. A., Hajra, A., Claxton, D. F., Marlton, P., Freedman, M., *et al.* (1993). *Science*, **261**, 1041.
15. Ledbetter, D. H., Rich, D. C., O'Connel, P., Leppert, M., and Carey, J. C. (1989) *Am. J. Hum. Genet.*, **44**, 2.
16. Kitsberg, D., Selig, S., Keshet, I., and Cedar, H. (1993). *Nature*, **366**, 588.
17. Bodrug, S. E., Burghes, A. H. M., Ray, P. M., and Worton, R. G. (1989). *Genomics*, **4**, 101.
18. Meitinger, T., Boyd, Y., Anand, R., and Craig, W. (1988). *Genomics*, **3**, 315.
19. Ballabio, A. and Andria, G. (1992). *Hum. Mol. Genet.*, **1**, 221.
20. Dauwerse, J. G., Jumelet, E. A., Wessels, J. W., Saris, J. J., Hagemeijer, A., Beverstock, G. C., *et al.* (1992). *Blood*, **79**, 1299.
21. Peake, I. R., Matthews, R. J., and Bloom, A. L. (1989). *Br. J. Haematol.*, **71**, 1.
22. Ketterling, R. P., Ricke, D. O., Wurster, M. W., and Sommer, S. S. (1993). *Hum. Mutat.*, **2**, 53.
23. Naylor, J., Brinke, A., Hassock, S., Green, P. M., and Gianelli, F. (1993). *Hum. Mol. Genet.*, **2**, 1773.
24. Lakich, D., Kazazian, H. H., Jr, Antonarakis, S. E., and Gitschier, J. (1993). *Nature Genet.*, **5**, 236.
25. Emery, A. E. H. (ed.) (1993). In *Oxford monographs on medical genetics Vol. 24: Duchenne Muscular Dystrophy*. Oxford University Press, Oxford.
26. Wijmenga, C., Hewitt, J. E., Sandkuijl, L. A., Clark, L. N., Wright, T. J., Dauwerse, H. G., *et al.* (1992). *Nature Genet.*, **2**, 26.
27. Van Deutekom, J. C. T., Wijmenga, C., Van Tienhoven, E. A. E., Gruter, A.-M., Hewitt, J. E., Padberg, G. W., *et al.* (1993). *Hum. Mol. Genet.*, **2**, 2037.
28. Timmerman, V., Nelis, E., Van Hul, W., Nieuwenhuijsen, B. W., Chen, K. L., Wang, S., *et al.* (1992). *Nature Genet.*, **1**, 171.
29. Boyce, F. M., Beggs, A. H., Feener, C. A., and Kunkel, L. M. (1991). *Proc. Natl. Acad. Sci. USA*, **88**, 1276.

30. Coffey, A. J., Roberts, R. G., Green, E. D., Cole, C. G., Butler, R., Anand, R., *et al.* (1992). *Genomics*, **12**, 474.
31. Verkerk, A. J. H. M., Pieretti, M., Sutcliffe, J. S., Fu, Y.-H., Kuhl, D. P. A., Pizzuti, A., *et al.* (1991). *Cell*, **65**, 905.
32. Bell, M. V., Hirst, M. C., Nakahori, Y., MacKinnon, R. N., Roche, A., Flint, T. J., *et al.* (1991). *Cell*, **64**, 861.

4

Cloning from PFGE-purified genomic DNA fragments and YACs

PAUL A. WHITTAKER and RAKESH ANAND

1. Introduction

An important step in the analysis of complex genomes is the construction of physical maps at the level of overlapping cloned DNA. Such cloned DNA maps provide both positional information and access to DNA fragments in a form that can be used for molecular analysis (e.g. probe isolation, sequencing, etc.). As a result, they have become an important tool in the structural and functional characterization of large genes, or the identification of disease genes by positional cloning. Preparative pulsed field gels, from which genomic DNA or YAC DNA is recovered for subsequent cloning, can facilitate the construction of these cloned DNA maps. Preparative gels can be used to

- enrich for large size genomic DNA fragments obtained by digestion with rare-cutting restriction endonucleases so that 'band-specific' subclones can be generated. These subclones can then be used to probe phage λ, cosmid, or YAC libraries for the isolation of overlapping clones;
- purify YAC DNA before subcloning into phage λ or cosmid vectors. This removes the need to distinguish clones containing inserts from the YAC from those containing yeast DNA;
- isolate a specific size fraction of DNA for phage λ or cosmid cloning as an alternative to sucrose gradients;
- concentrate or 'focus' a desired size range of DNA fragments in a small area whilst eliminating smaller fragments. This step is important in YAC cloning to reduce the number of YACs with small inserts, and eliminate circularized vector and self-ligated vector arms.

Although the loading capacity of pulsed field gels is not as high as that of sucrose gradients, they give more control over the sizes of DNA fractionated and, with care, allow extremely large DNA fragments to be isolated.

In this chapter we concentrate on the use of preparative pulsed field gels for the generation of band-specific probes and YAC sublibraries. The methods presented should serve as a starting point for workers engaged in the detailed analysis of large regions of genomic DNA. However, the protocols are not sacrosanct and may, in some cases, need to be tailored to the individual project and the experience of the experimenter(s).

Methods for 'focusing' of DNA fragments above a desired size range for the construction of YAC libraries can be found in Chapter 5.

2. General considerations

2.1 Precautions

In order to isolate large DNA fragments by preparative PFGE it is essential to use reagents of the highest purity throughout the process and to guard against nuclease contamination or shear damage to the DNA. Points to note are the following.

- Agarose used for plug preparation and preparative gels should be checked for its compatibility with subsequent enzyme reactions. SeaPlaque low melting temperature (LMT) agarose (from FMC) is reliable and works well for both applications. However, due to occasional batch to batch variation, it may be desirable to test a few batches to find one which yields the most intact DNA in plugs, and has no restriction enzyme inhibitors present. This can be done by embedding λ DNA in the agarose and testing for degradation and digestion. Alternatively, use pre-tested 'Insert' agarose from FMC.
- As preparative gels are made with LMT agarose, run times and switch intervals should be adjusted to compensate for the slightly faster mobility.
- The gel box, casting assembly, and other implements should be cleaned carefully and gloves should be worn to prevent nuclease contamination.
- To prevent DNA damage, DNA to be used for cloning should not be stained with ethidium bromide or exposed to UV light.

2.2 Cloning of genomic fragments

Important points to note:

- Use of a somatic cell hybrid containing the chromosome of interest as the starting material provides an enrichment step as well as facilitating the screening of resulting clones.
- The restriction enzymes used should generate a gel profile such that the fragment of interest is located outside the bulk of the genomic DNA fragments. If this is not possible using single or double restriction digests, sequential PFGE fractionation may be necessary.
- It is essential to have a strategy to identify readily probes derived from the PFGE-fractionated large DNA fragment.

3. Preparation of DNA samples

3.1 Genomic plugs

The rate of migration of DNA through a pulsed field gel can be significantly retarded by high DNA concentration in the original plug. This can result in a lack of correlation between the size markers and the actual size of fractionated DNA fragments, and can also lead to trapping of small DNA in the larger size fractions. *Protocol 1* produces DNA plugs at 90 μg/ml and we suggest that the starting DNA concentration in plugs should never exceed 180 μg/ml.

Protocol 1. Preparation of genomic DNA plugs

Equipment and reagents

- Dulbecco's saline: 0.2 g/litre KCl, 0.2 g/litre KH$_2$PO$_4$, 8.0 g/litre NaCl, and 2.16 g/litre Na$_2$HPO$_4$.7H$_2$O. Adjust the pH to 7.4 with HCl; autoclave to sterilize
- Low melting temperature (LMT) agarose (FMC, SeaPlaque): 1% (w/v) solution in Dulbecco's saline equilibrated at 37–40°C
- NDS: prepare this solution by dissolving 93 g EDTA in 350 ml H$_2$O. Add 0.605 g Tris base and solid NaOH pellets until the pH is above 8.0. Add 5 g lauroyl sarcosine dissolved in 50 ml distilled H$_2$O. Adjust the pH to 9.5 using concentrated NaOH and make the volume up to 500 ml. Filter through a 0.2 μm filter, or autoclave, taking care to avoid excessive frothing, and store at 4°C. This solution can be replaced with YLB (*Protocol 2*)
- 50 ml sterile polypropylene conical centrifuge tube (or similar)

- 20 mg/ml pronase (Boehringer Mannheim) in NDS. Incubate at 37°C for 30 min to digest any contaminants before diluting to 1 mg/ml with NDS. Proteinase K is an expensive alternative.
- Perspex plug mould available from several suppliers of PFGE kit. The average plug size is 6 × 2 × 10 mm. Before use, clean the mould by boiling in 0.25 M HCl for 10 min and then wash it several times in distilled water to remove all traces of the acid. Cover one side of the mould with tape and precool it on ice.
- Water baths or incubators at 37°C and 50°C
- Light microscope and haemocytometer
- Centrifuge and carriers capable of 500 *g*
- Micropipettes and tips
- Bent glass Pasteur pipette or folded pH paper dip-stick (Whatman)

Method

1. Harvest the cells by centrifugation (500 *g*, 10 min), wash in Dulbecco's saline, and suspend in 1 ml of the same buffer. Disrupt clumps of cells by sucking backwards and forwards through a micropipette tip. Remove an aliquot and count the number of cells using a haemocytometer. Dilute the cell suspension to 3×10^7 cells per ml, warm to 37°C, and hold at this temperature.

2. Add an equal volume of 1% (w/v) LMT agarose at 37–40°C and mix.

3. Dispense the mixture into the Perspex plug mould and leave the agarose to set.

4. Remove the tape from the mould and gently push the plugs out using a bent glass Pasteur pipette or a folded pH paper dip-stick into a 50 ml polypropylene conical centrifuge tube containing 2–4 volumes of 1 mg/ml pronase in NDS.

71

Protocol 1. *Continued*

5. Incubate at 50°C overnight.

6. Replace the pronase solution with 2–4 volumes of fresh pronase solution and continue incubation at 50°C for another 24 h.

7. Rinse the plugs twice for 2 h in NDS without pronase and store in NDS at 4°C. Plugs are stable for at least a year.

Procedures for harvesting and washing tissue culture cells, white blood cells, sperm, and cells from fresh or frozen tissue can be found in ref. 1.

3.2 YAC plugs

The method detailed in *Protocol 2* gives plugs at a concentration of $\sim 3 \times 10^8$ cells per plug, equivalent to ~ 5 μg yeast DNA and ~ 65 ng, 130 ng, and 260 ng of a 200 kb, 400 kb, and 800 kb YAC respectively. An important point to note is that extensive washing of the plugs in YLB is essential to remove diffusible material from the plugs which forms a background smear on pulsed field gels. Lyticase treatment of the yeast cells before embedding in LMT agarose increases the efficiency of spheroplasting. Zymolyase 20T (ICN) can be used as a substitute for lyticase.

Protocol 2. Preparation of YAC DNA plugs

Equipment and reagents

- SD medium: 7 g/litre Bacto yeast nitrogen base without amino acids, 20 g/litre glucose, 55 mg/litre adenine and tyrosine, 14 g/litre casamino acids; autoclave to sterilize
- 50 mM EDTA pH 8.0; autoclave to sterilize
- YRB: 1 M sorbitol, 10 mM Tris–HCl pH 7.5, 20 mM EDTA; autoclave to sterilize and add β-mercaptoethanol to 14 mM (1 μl of 14 M stock per ml of YRB) prior to use
- Lyticase (Sigma): 10 000 units/ml in sterile 50 mM Tris–HCl pH 7.5, 1 mM EDTA, 10% (v/v) glycerol
- Low melting temperature (LMT) agarose (SeaPlaque, FMC): 1% (w/v) solution in YRB equilibrated at 37°C

- YLB: 0.1 M EDTA, 10 mM Tris–HCl pH 8.0, 1% (w/v) lithium dodecyl sulphate; filter and sterilize or autoclave, avoiding excessive frothing
- 10% (w/v) SDS
- Perspex plug mould (see *Protocol 1*)
- Water baths or incubators at 37°C and 50°C
- Light microscope and haemocytometer
- Incubator shaker at 30°C
- 50 ml sterile polypropylene conical centrifuge tubes (or similar)
- Centrifuge and carriers capable of 500 *g*
- Micropipettes and tips

Method

1. Inoculate 2 × 500 ml of SD medium with the appropriate YAC clone and shake overnight (18–24 h) at 30°C.

2. Harvest the cells by centrifugation (500 *g*, 10 min) and wash twice with 50 mM EDTA pH 8.0.

3. Resuspend the cell pellet in YRB at a concentration of 5×10^9 cells/ml and add lyticase to a final concentration of 50–100 units/ml.

4. Incubate the suspension at 37°C for 1 h.

5. Check for spheroplast formation by removing 1 μl aliquots into 100 μl YRB and 100 μl 10% (w/v) SDS and counting the number of spheroplasts in each using a haemocytometer. Approximately 90% of cells have usually spheroplasted by this time. If not, add more lyticase and continue the incubation.

6. Gently mix the suspension with an equal volume of 1% LMT agarose and dispense into the Perspex plug mould.

7. Once set, push the plugs out into 2 volumes of YLB and gently shake at room temperature for 2 h.

8. Replace the solution with 10 volumes of YLB and incubate at 50°C overnight.

9. Repeat step 8.

10. Rinse the plugs once in YLB and store in YLB at room temperature. Plugs are stable for at least a year.

4. Manipulation of DNA samples

4.1 Genomic plugs

Residual pronase or Proteinase K in the plugs is removed by extensive washing in TE buffer (*Protocol 3*) before restriction enzyme digestion. The protease inhibitor PMSF (which is toxic and unstable) can be included in the first wash at a concentration of 0.1 mM, but this is not usually necessary unless difficulties with restriction enzyme digestion are encountered. As a rule, more enzyme is needed to digest DNA in agarose than in liquid. New England Biolabs has tested the minimum amount of enzyme required for complete digestion of purified λ DNA and adenovirus-2 DNA embedded in agarose and lists the results in the reference appendix of its latest catalogue. These results can be used as a guide, but it should be noted that more enzyme is needed to digest mammalian genomic DNA than purified viral DNA. In addition, the purity of the substrate DNA can significantly alter its digestibility, as can the source of the restriction enzyme.

Longer digestion times can be used to decrease the amount of restriction enzyme needed for complete digestion. However, a number of enzymes (e.g. *Sma*I) are not stable for longer than one hour at 37°C, therefore a second aliquot of enzyme is usually added after the initial incubation. Information on the activity and stability of all restriction enzymes during extended incubation can be obtained from the commercial suppliers.

Nuclease contamination of genomic plugs can produce DNA smears that resemble restriction digests. Consequently each new batch of plugs should be tested for nuclease activity before restriction analysis by incubating a plug in restriction enzyme buffer at 37°C without restriction enzyme and then comparing with an untreated plug on a pulsed field gel. Most of the DNA should remain in the gel slot with only a small amount of material running from the slot to the compression zone. If degradation is detected, plugs should be incubated for longer in pronase in NDS (*Protocol 1*). For cases where nuclease activity is suspected in the restriction buffer used, each component of the buffer system should be checked. BSA is a common culprit. If problems are encountered, autoclaved gelatin can be substituted. Spermidine may also be added as it aids digestion with some enzymes. If used, the final concentration must not exceed 2 mM in buffers with 50–100 mM salt and 5 mM in buffers with ≥ 100 mM salt.

Protocol 3. Restriction enzyme digestion of genomic plugs

Equipment and reagents

- TE buffer: 10 mM Tris–HCl pH 7.5, 1 mM EDTA
- Restriction enzyme and 10 × restriction digest buffer
- BSA (10 mg/ml) or gelatin (Sigma): 1 mg/ml in sterile distilled water, autoclaved, and stored at −20°C

- Spermidine: 0.1 M made up in sterile distilled water, filtered through a 0.2 μm filter, and stored at −20°C
- Stop buffer: 0.5 × electrophoresis buffer (see *Protocol 4*), 10 mM EDTA, and 0.25% (w/v) orange-G dye
- Micropipettes and tips

Method

1. Wash the plugs with agitation for at least 3 × 30 min in 10 volumes of cold TE buffer on ice (ideally, the first wash should be overnight. 0.1 mM phenylmethylsulfonyl fluoride (PMSF) can be included if desired).

2. Equilibrate the plugs in 10 volumes of 1 × restriction buffer for 30 min on ice.

3. Replace the buffer with 2 volumes of 1 × restriction buffer containing 100 μg/ml BSA (or gelatin) and 2–5 mM spermidine (optional).

4. Add the appropriate amount of enzyme and incubate for 2 h to overnight at 37°C (or the recommended temperature for the enzyme).

5. After digestion, remove all the restriction buffer and add 3 volumes of stop buffer.

6. Store at 4°C ready for loading on to a pulsed field gel. For longer term storage, plugs can be stored in NDS, but they will need to be reequilibrated with stop buffer before loading on a gel.

4.2 YAC plugs

To prepare plugs for PFGE, dialyse against 20–50 volumes of stop buffer (*Protocol 3*) for about 1 h before loading on the gel. Ideally, one overnight wash in about 50 volumes of TE buffer (*Protocol 3*) followed by equilibration in approximately 10 volumes of stop buffer ensures no residual salt effects during subsequent electrophoresis.

5. Electrophoresis

5.1 Introduction

A number of pulsed field systems have been described since the original work of Schwartz and Cantor in 1984. The differences between these systems reside in the electrode geometry used, field homogeneity, and the method of re-orientation of the electric fields (Chapter 1). Important considerations for preparative work are

- whether the apparatus gives straight tracks;
- the resolution within a particular size range; and
- how large a portion of the gel provides useful separation.

We use the crossed-field system (2), which permits large gels to be run with straight lanes. However, any PFGE system with comparable characteristics can be used.

5.2 Casting the gel and sample loading

Although ethidium bromide is often incorporated into conventional gels, it should not be incorporated into pulsed field gels because it retards DNA migration and also makes the DNA susceptible to light-induced nicking. For preparative gels, a 1% (w/v) low melting temperature agarose (SeaPlaque, FMC) separating gel is cast within a 1.5% (w/v) agarose (Sigma type 1) surround (*Figure 1*) to make subsequent handling easier. *Protocol 4* describes preparation of the latter.

Protocol 4. Preparation of molten agarose

Equipment and reagents

- Water (glass distilled, preferably double distilled)
- Agarose (Sigma type 1, or equivalent)
- Conical flask (should hold ~2.5 times the volume of the agarose solution)
- Microwave oven (650 W)

- 0.5 × Tris-acetate electrophoresis (TAE) buffer: 0.02 M Tris-acetate, 0.0005 M EDTA. To prepare 1 litre of 50 × concentrated stock solution, mix 242 g Tris base, 57.1 ml glacial acetic acid, and 100 ml 0.5 M EDTA pH 8.0 with water to 1 litre.

Method

1. Add the desired amount of agarose to 0.5 × TAE buffer in a conical flask.

2. Weigh the conical flask and note the weight.

3. Heat the flask for 2–3 min in the microwave oven on full power.

Protocol 4. *Continued*

4. Remove the flask and gently swirl to resuspend the agarose particles (**Caution**: microwaved solutions can boil over suddenly when swirled).

5. Return to the microwave and heat using a medium setting for at least 2 min more.

6. Reweigh the flask and bring it back to its original weight by adding water.

7. Cool to ~45°C before pouring the gel.

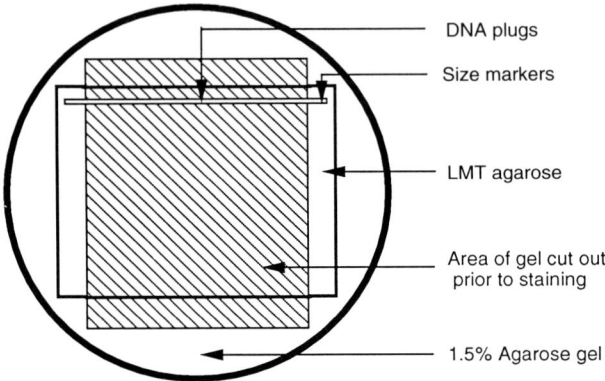

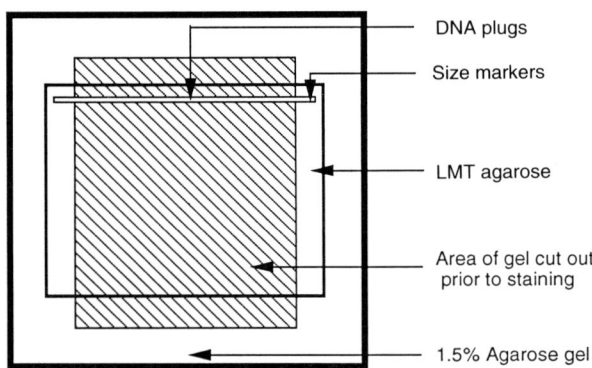

Figure 1. Diagrammatic representation of preparative pulsed field gels showing a 1.5% (w/v) support gel with a 1% (w/v) LMT agarose insert and a single long sample-loading well. Upper panel is for systems using circular gels (e.g. Hula gel (Hoefer Scientific Instruments), rotating field gel electrophoresis (RFGE)), whereas the lower panel is for systems using square or rectangular gels (e.g. contour clamped homogeneous electric field (CHEF), field inversion gel electrophoresis (FIGE)). The shaded area is the region cut out prior to staining and UV photography.

Make the sample loading slot in the gel using a single long-slot comb. Flood the loading slot with electrophoresis buffer and load intact plugs containing digested genomic DNA or undigested YAC DNA into the sample well with size markers at either end. Once all the plugs are loaded, remove excess buffer using a tissue and seal the plugs in place using molten LMT agarose (*Figure 1*).

5.3 Electrophoresis conditions

The size range of DNA fragments that can be separated on pulsed field gels is dependent on the pulse time, voltage, temperature, and electrophoresis buffer (see Chapter 1). Therefore, separation conditions should be chosen to optimize band sharpness and resolution (the ability to discriminate between two closely migrating bands).

5.4 Processing of preparative gels

5.4.1 Genomic samples

After electrophoresis, cut out about 90% of the area of the LMT agarose gel containing fractionated DNA (shaded area, *Figure 1*) leaving approximately half of one digested plug width and size markers on either side. Store this cut-out LMT agarose slab in electrophoresis buffer. Meanwhile, fill the cut-out area of the preparative gel with agarose solution to maintain the shape and integrity of the gel, stain it, and photograph with a ruler placed on the side of the gel. This photograph and the image of the ruler within it is used to measure the distance from the loading slot that the required size DNA fragments have migrated, i.e. if the fragment to be purified is 600–700 kb and the size markers indicate that it should have migrated 7–9 cm from the loading slot, then that is the target region. Using accurate measurements, cut out the target region from the stored LMT gel and slice it into 2 mm strips perpendicular to the direction of electrophoresis. Number the strips sequentially and store them individually in sealed tubes or wrapped in cling film (wrap and seal) at 4°C. Then transfer the reconstructed stained gel DNA to a nylon membrane and hybridize with the appropriate probe to identify the exact distance of the required (hybridizing) band from the loading slot. This distance can be used to identify the gel slice(s) containing the required DNA fragment. A diagrammatic example is shown in *Figure 2*. Confirmation of the DNA identity in the gel slice(s) can be obtained either by running small amounts of the DNA in the agarose strips on a conventional gel, blotting, and probing (3), or by hybridizing dot blots of the DNA (4, 5).

An alternative, which works equally well, particularly for bands > 500 kb in size (because of minimal diffusion of DNA), is to store the cut-out gel region (shaded area, *Figure 1*) in a humidified, sealed container, or wrapped in cling film in the fridge. Following autoradiography, the required DNA region can be recovered by placing the stored LMT gel on top of a glass plate

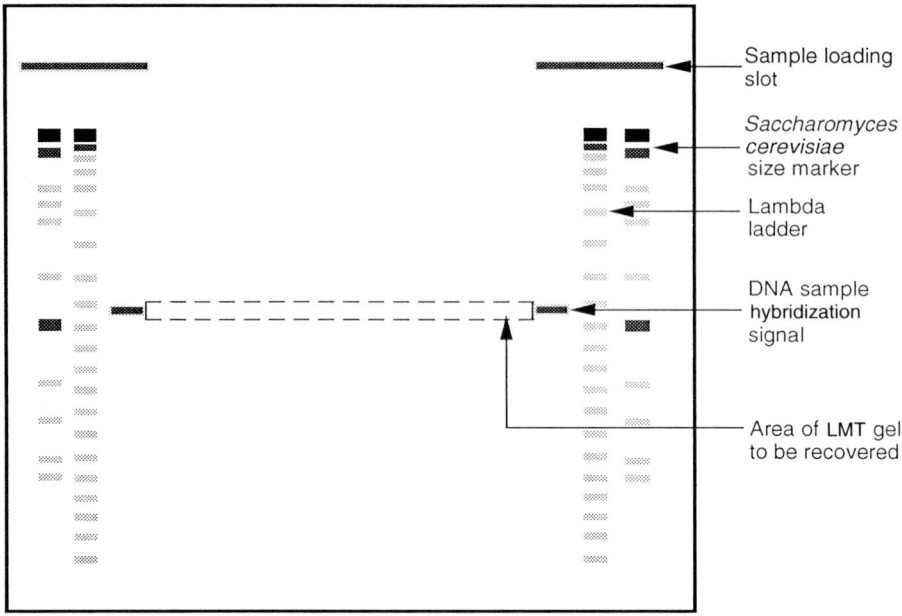

Sample loading slot

Saccharomyces cerevisiae size marker

Lambda ladder

DNA sample hybridization signal

Area of **LMT** gel to be recovered

Figure 2. Diagrammatic representation of an autoradiograph showing the signal from the ends of the sample-loading slot which can be used to measure the exact distance of migration of the hybridizing band from the loading slot to identify the slice number(s) containing the target DNA. Alternatively, the loading slot signal on the autoradiograph can be aligned with the loading slot of the stored LMT gel to recover the region containing the target DNA. Use of small amounts of radiolabelled yeast and lambda DNA during hybridization enables the size markers to be detected on the autoradiograph and also ensures that the loading slot is clearly visible.

and the autoradiograph under the glass plate. Once the sample loading slot and its image on the autoradiograph is aligned, the region of gel containing the required DNA can be recovered by slicing it out of the LMT gel (*Figure 2*).

5.4.2 YAC samples

After electrophoresis, slice off the two end lanes containing the size markers and half one slot plug width of the central long slot and stain with ethidium bromide. Mark the position of the YAC on the stained portion of the gel by notching the gel strip with a scalpel blade. Then place the stained and unstained pieces of the gel side-by-side and excise the YAC from the unstained central region of the gel. To ensure accuracy, place a piece of graph paper under the glass plate containing the gel so that you have a guide to follow when cutting out the strip. This also helps prevent movement of the gel, a problem encountered when using a ruler as a guide because of deformation of the gel due to downward pressure. Stain the unstained portion of the gel to confirm the accuracy of the band excision. Store the excised band in 0.5 × TAE at 4°C until required.

6. Recovery of DNA

For cloning experiments, a rule of thumb is that the size of the starting DNA should be at least 3–5 times the size of the desired restriction fragments to be cloned to minimize the number of DNA fragments with one sheared end and one restriction-enzyme generated end. Many methods for the recovery of DNA from standard agarose gels (e.g. electroelution on to DEAE cellulose, GeneClean purification, etc.) are unsuitable for large DNA. However, it is possible to use these methods if the DNA is reduced in size by digesting with a restriction enzyme prior to purification (*Protocol 5*). This method is useful when complete digest products are to be cloned using plasmid or phage insertion vectors. Alternatively, if the DNA is to be partially digested for cloning into a cosmid or phage λ replacement vector, it can be released from the agarose by digestion with β-agarase 1 (*Protocol 6*), an enzyme from *Pseudomonas atlantica* which digests polymerized agarose to carbohydrate molecules which will no longer gel and do not interfere with subsequent DNA manipulations (restriction digestion, ligation, and transformation). The enzyme is more effective on melted agarose than on solid agarose, so the LMT agarose slices containing the DNA are first melted at 65°C before treatment with agarase at 37–40°C. The enzyme has an optimum activity at pH 6.5, but greater than 75% of this activity is maintained from pH 5.0–8.5. Sodium chloride or EDTA concentrations between 50 and 500 mM have no effect on agarase activity so the enzyme is active in a variety of buffers including commonly used electrophoresis buffers (TAE, TBE). However, the optimum buffer is 10 mM bis-Tris–HCl pH 6.5, 1 mM EDTA, giving twice and ten times the activity observed in TAE and TBE buffers, respectively. Carbohydrates and β-agarase 1 can be removed by ethanol precipitation, or phenol extraction followed by ethanol precipitation. However, these manipulations may shear the DNA and therefore subsequent digestion of the DNA is usually carried out in the agarase-treated agarose. If required, the DNA can be concentrated using 'microcon' microconcentrators from Amicon. DNA prepared by this method is a more suitable substrate for partial digestion.

An important point to note is that it has been shown that polyamines (6, 7) should be present whenever agarose containing high molecular weight DNA is melted at 68°C, to protect DNA from degradation.

Protocol 5. Recovery of DNA fragments—method A

Equipment and reagents

- Restriction enzyme and 10 × digest buffer
- GeneClean kit (Bio 101)
- Water baths or heating blocks at 37°C and 65°C
- Dephosphorylated vector DNA (*Protocol 7*)
- Microcentrifuge
- 100 × polyamines solution: 30 mM spermine, 75 mM spermidine; solution made using sterile distilled water, filtered, sterilized, and stored at −20°C
- 1.5 ml microcentrifuge tubes
- Micropipettes and tips

Protocol 5. *Continued*

Method

1. Cut the LMT slice into 5–10 small pieces and equilibrate in a ten-fold excess of 1 × restriction digest buffer supplemented with 1 × polyamines.

2. Transfer the agarose pieces to a 1.5 ml microcentrifuge tube and heat to 65°C for 10 min to melt the agarose.

3. Cool the molten agarose to 37°C and add restriction enzyme to 100 units/ml.

4. Incubate at 37°C for 2 h.

5. Add 0.5 µg of dephosphorylated vector and purify total DNA using GeneClean and the protocol supplied with the kit.

Protocol 6. Recovery of DNA fragments—method B

Equipment and reagents

- β-agarase 1 (New England Biolabs, Calbiochem, or Boehringer Mannheim)
- Agarase buffer 10 mM bis-Tris–HCl pH 6.5, 1 mM EDTA
- 100 × polyamines solution: 30 mM spermine, 75 mM spermidine; solution made using sterile distilled water, filtered, sterilized, and stored at −20°C
- Centricon 10 microconcentrator (Amicon)
- Water bath at 65°C
- Centrifuge with fixed angle rotor adaptors or carriers accepting 17 × 100 mm tubes and capable of 5000 *g*
- Micropipettes and tips

Method

1. Determine the volume of the slice by weighing in a tared sterile tube.

2. Equilibrate the gel slice in two volumes 10 mM bis-Tris–HCl pH 6.5, 1 mM EDTA supplemented with 1 × polyamines solution on ice for 30 min twice.

3. Melt the agarose pieces at 65°C for 10 min.

4. Cool to 37°C and add agarase according to the manufacturers' recommendations.

5. Incubate at the temperature recommended by the agarase vendor for 1 h.

6. If required, the sample can be concentrated by loading the mix into a Centricon 10 microconcentrator and centrifuging at 5000 *g* for up to 1 h depending on the initial sample size and the degree of concentration required.

7. The agarased DNA is stable at 4°C for several weeks.

7. Cloning of purified DNA

7.1 Choice of vector

Plasmid, phage λ, and cosmid vectors are all suitable for the cloning of PFGE-purified DNA. The main point to consider in the choice of vector is the degree of coverage of the purified fragment required in the resulting library, i.e. is a collection of sequentially overlapping clones (a contig) covering the entire region required, or a series of probes? This in turn is affected by the source of the purified fragment: genomic DNA, or YAC DNA. Obviously, YAC DNA is the preferred source for the preparation of a contig covering a given region because of the larger amounts of material that can be purified, whereas both sources are suitable for the generation of probes.

For the generation of contigs, cosmid and phage λ replacement vectors are more suitable because of their greater cloning capacity (30–45 kb and 10–25 kb, respectively). In addition, a number of vectors have been constructed with features which facilitate the identification and alignment of sets of overlapping clones (8). Both cosmid and phage λ vectors allow efficient construction of libraries from small amounts of DNA, but the latter are the easiest to screen by hybridization because phage plaques contain large amounts of unpackaged DNA along with phage. Also, the smaller insert size of λ vectors imposes a less stringent requirement on the size of the purified DNA. Cosmid vectors should only be used for cloning if the size of the starting YAC is in excess of 200 kb.

For the generation of probes from pulsed-field-purified DNA fragments, both plasmid and phage λ insertion vectors are the vectors of choice. Plasmid vectors work well with DNA fragments of small size (<10 kb) and simple structure (palindromic sequences are difficult to clone). Moreover, the transformation efficiencies that can now be obtained using competent *E. coli* cells ($\geq 10^8$ colonies/μg of supercoiled plasmid DNA), or electrotransformation of *E. coli* cells (10^9–10^{10} colonies/μg of supercoiled plasmid DNA) rivals that obtained by *in vitro* packaging of phage λ DNA (10^8–10^9 plaques/μg DNA). The main disadvantage of plasmid vectors is that they show a bias towards cloning small inserts, which can mean that where clones need to be identified on the basis of their content of specific interspersed repetitive sequences (e.g. the cloning of human DNA fragments from hybrid cell lines), some clones may go undetected. A suitable plasmid vector would be pBluescript II from Stratagene because it has a number of features designed to facilitate the analysis of cloned inserts.

Phage λ insertion vectors show less of a size bias and because ligated concatemers are favoured substrates for *in vitro* packaging, a large molar excess of λ DNA can be used in the ligation reaction to increase the cloning efficiency and reduce co-ligation of non-contiguous insert fragments. Although a number of insertion vectors have been described (9), the λZAP vector

(Stratagene; (10)) is particularly useful because cloned inserts can be excised and converted into plasmids *in vivo*.

7.2 Vector preparation

Because of the small amounts of DNA to be cloned, it is important to reduce the level of non-recombinant plasmids, phage, or cosmids caused by self ligation of the vector molecules. The simplest and most effective way to do this with plasmid and phage λ insertion vectors is to dephosphorylate the vector DNA using calf intestinal alkaline phosphatase (CIAP; *Protocol 7*). Points to note about the use of phosphatase are given below.

- Restriction endonuclease cleavage and dephosphorylation of the vector DNA can be performed simultaneously. Add spermidine to assist restriction enzyme cleavage as well as help maintain the integrity of the cleaved ends.

- It is critical to inactivate and remove the CIAP before proceeding. Achieve this either by heat inactivating the enzyme in the presence of EGTA and removing the CIAP by phenol extraction, or by digestion with Proteinase K in the presence of EDTA, followed by phenol extraction.

- Efficiency of dephosphorylation should be checked by attempting to ligate the CIAP-treated vector to itself. Check the integrity of the dephosphorylated ends by ligating CIAP-treated vector DNA in the presence of untreated DNA (e.g. λ DNA fragments with compatible ends). Monitor these reactions by agarose gel electrophoresis: dephosphorylated vector DNA should only show a shift in mobility when ligated to DNA that retains its 5'-phosphate group. A precise estimate of the efficiency of dephosphorylation can be achieved by comparing the transformation efficiency (plasmid or cosmid vectors), or packaging efficiency (phage λ vectors) of CIAP-treated and ligated DNA with untreated and ligated DNA.

Protocol 7. Dephosphorylation of vector DNA with calf intestinal alkaline phosphatase

Equipment and reagents

- Restriction enzyme and 10 × digest buffer
- BSA (10 mg/ml)
- Spermidine: 0.1 M made up in sterile distilled water, filtered through a 0.2 μm filter, and stored at −20°C
- Calf intestinal alkaline phosphatase (CIAP; Boehringer Mannheim, molecular biology grade)
- 0.5 M EGTA pH 8.0
- Phenol: containing 0.1% (v/v) hydroxyquinoline and equilibrated with 0.1 M Tris–HCl pH 8.0 containing 0.2% (v/v) β-mercaptoethanol
- Chloroform:isoamyl alcohol (24:1)
- 5 M ammonium acetate
- 100% ethanol
- TE buffer (*Protocol 3*)
- Water baths or heating blocks at 37°C or 65°C
- Microcentrifuge
- 1.5 ml microcentrifuge tubes
- Micropipettes and tips
- Mini-gel apparatus

Method

1. Digest the vector DNA (10–20 μg) with a 2–3 fold excess of the desired restriction enzyme in the appropriate buffer containing 100 μg/ml BSA and 5 mM spermidine for 60 min at 37°C. Remove an aliquot (0.2 μg) and analyse on an agarose mini-gel. If digestion is incomplete, add more enzyme and continue the reaction.

2. When digestion is complete, add CIAP (1 unit/50 pmoles of ends) and continue the incubation at 37°C for 30 min.

3. Add one tenth volume of 0.5 M EGTA pH 8.0 and incubate at 65°C for 60 min.

4. Extract with an equal volume of phenol and chloroform:isoamyl alcohol twice, and an equal volume of chloroform:isoamyl alcohol once. Remove the aqueous phase to a fresh tube after each extraction.

5. Ethanol precipitate the DNA for 10 min at room temperature after adding an equal volume of 5 M ammonium acetate and two volumes of ethanol.

6. Recover the DNA by centrifugation in a microcentrifuge at 12 000 g for 15 min.

7. Dry the pellet under vacuum and resuspend at a concentration of 1 mg/ml in TE buffer.

The use of dephosphorylated vector necessarily precludes the use of CIAP-treated donor DNA. This does not present a problem when the purified DNA is cloned into plasmid or λ insertion vectors after limit digestion with the appropriate enzyme, because co-ligation of non-contiguous insert fragments will be easily detected due to the presence of more than one restriction fragment in the resulting clone. However, if the purified DNA is partially digested before cloning, co-ligation of non-contiguous sequences can present problems because it is difficult to detect in the resulting clones. A population of DNA fragments of an average size of 20 kb will contain many molecules that are shorter than the mean. Although these shorter molecules may compose only a fraction of the total *weight* of the DNA, they make a substantial contribution to the *number* of molecules in the population. Since the kinetics of ligation are determined by the concentration of reactive termini of DNA, these smaller molecules can become incorporated into concatemers. This could be avoided by careful size fractionation of the partially digested DNA using agarose gel electrophoresis, but this method is unsuitable for nanogram quantities of DNA. Therefore, the partially digested DNA must be rendered incapable of self ligation by enzymatic means. In the case of cosmid vectors, the partially digested DNA can be dephosphorylated using CIAP and cloned into cosmid 'arms' prepared using a multiple enzyme digestion procedure ((11); *Protocol 8*). In the case of phage λ vectors, the 5′ protruding ends of

DNA partially digested with *Bam*HI, *Mbo*I, *Bgl*II, or *Sau*3AI can be partially filled in with dGTP and dATP. 5′-GA overhangs result and the insert DNA will not self ligate, preventing multiple inserts. However, the DNA will ligate to vector DNA digested with *Xho*I or *Sal*I and partially filled in with dTTP and dCTP to generate 5′-TC overhangs ((12); *Protocol 9*).

Although cosmid 'arms' can be prepared using vectors containing only one *cos* site, the procedure is easier with vectors containing double *cos* sites, because the number of enzymatic manipulations is decreased and the arms do not need to be purified by preparative gel electrophoresis. A number of double *cos* site vectors have been constructed. The preparation of 'arms' using one of these, vector SuperCos-1 (available from Stratagene), is detailed in *Protocol 8*. This vector has several features which facilitate overlap detection between clones and restriction mapping of cloned inserts. It also carries the neomycin resistance marker so that cosmids can be transferred into mammalian cells.

The partial fill-in method can be used with any vectors containing suitably placed *Xho*I or *Sal*I sites. However, to simplify downstream analysis of clones, vectors designed to facilitate the detection of overlapping clones and the restriction mapping of cloned inserts are recommended. Suitable vectors are λFIX II and λDASH II from Stratagene; λGEM-11 and -12 from Promega; and λEMBL3*cos*W (13). Since λEMBL3*cos*W has been used for the construction of YAC sublibraries (14), the preparation of this vector is given in *Protocol 9*. Although partial fill-in of the vector reduces the level of non-recombinant phage substantially, in order to reduce the level close to zero, it is necessary to cleave the vector with a second enzyme and to dephosphorylate the doubly cleaved, filled-in DNA. Cleavage with a second enzyme helps for two reasons:

- non-recombinant plaques can result because a small proportion of vector molecules will not cleave with *Xho*I, due to the accumulation of mutants during growth of the phage. Cleavage with a second enzyme will render these vector molecules inactive. This problem can also be minimized by plaque-purifying the vector frequently;

- it yields vector arms that are compatible with the partially cleaved DNA to be cloned, but a 'stuffer' fragment with incompatible termini. The short segment of the polycloning site carrying *Xho*I and *Eco*RI termini is removed by precipitation with isopropanol.

Protocol 8. Preparation of SuperCos-1 cosmid vector arms

Equipment and reagents

- Vector SuperCos-1 DNA (Stratagene)
- Restriction enzymes *Xba*I and *Bam*HI and their 10 × reaction buffers
- BSA (10 mg/ml)
- Spermidine (see *Protocol 7*)
- CIAP (see *Protocol 7*)

- 0.5 M EGTA pH 8.0
- Phenol (see *Protocol 7*)
- Chloroform:isoamyl alcohol (24:1)
- 100% ethanol
- 5 M ammonium acetate
- TE buffer (*Protocol 3*)

- Water baths or heating blocks at 37°C and 65°C
- Microcentrifuge
- 1.5 ml microcentrifuge tubes
- Micropipettes and tips
- Mini-gel apparatus

Method

1. Digest 20 μg SuperCos-1 DNA with 40 units of *Xba*I in a reaction volume of 100 μl for 2 h at 37°C.

2. Analyse 1 μl (0.2 μg) of the reaction by electrophoresis through a 0.8% (w/v) agarose mini-gel to check that the digest is complete (include 0.2 μg of undigested vector for comparison). If supercoiled or nicked circular forms of the vector are visible, add an additional 20 units of *Xba*I and continue the incubation for a further 1–2 h.

3. When the digestion is complete, dephosphorylate and process the vector as described in steps 2–7 of *Protocol 7*.

4. Digest the *Xba*I-digested, dephosphorylated vector with 40 units *Bam*HI, as described in step 1, and analyse a 1 μl aliquot as described in step 2. Only two vector bands should be visible; no linear plasmid should remain. If digestion is incomplete, add additional enzyme (step 2).

5. When digestion is complete, extract the DNA with phenol and recover the DNA as described in steps 4–7 of *Protocol 7*.

Protocol 9. Preparation of λEMBL3*cos*W *Xho*I half-site arms

Equipment and reagents

- Restriction enzymes *Xho*I and *Eco*RI and their reaction buffers
- BSA (10 mg/ml)
- Spermidine (see *Protocol 7*)
- Klenow fragment of *E. coli* DNA polymerase I
- 10 mM dCTP and 10 mM dTTP
- TE buffer (*Protocol 3*)
- Phenol (see *Protocol 7*)
- Chloroform:isoamyl alcohol (24:1)
- 5 M ammonium acetate

- 100% isopropanol
- 100% ethanol
- CIAP (see *Protocol 7*)
- 0.5 M EGTA pH 8.0
- 0.1 M EDTA pH 8.0
- Water baths or heating blocks at 37°C and 65°C
- Microcentrifuge
- 1.5 ml microcentrifuge tubes
- Mini-gel apparatus
- Micropipettes and tips

Method

1. Digest 20 μg of λEMBL3*cos*W DNA for 2 h at 37°C with 40 units *Xho*I in a reaction volume of 100 μl as described in step 1 of *Protocol 7*.

Protocol 9. *Continued*

Analyse 1 μl on a 0.5% (w/v) agarose mini-gel to confirm that the digest is complete. Add another 20 units of enzyme and digest for a further 2 h if uncut vector is visible.

2. When the digest is complete, add one-tenth volume each of 10 mM dCTP and 10 mM dTTP, 20 units of Klenow fragment, and incubate for 30 min at 37°C.

3. Add EDTA pH 8.0 to 20 mM. Extract the DNA with phenol and recover the DNA as described in steps 4–7 of *Protocol 7*.

4. Digest the *Xho*I-digested, partially filled-in vector DNA with 40 units *Eco*RI in a reaction volume of 100 μl for 2 h at 37°C.

5. Add 1 unit CIAP and continue the incubation for 30 min at 37°C.

6. Process the DNA as described in steps 3–7 of *Protocol 7*, except precipitate the DNA by adding 0.6 volumes of isopropanol instead of 2 volumes of ethanol.

Monitor the success of the fill-in reaction by attempting to self-ligate aliquots of the vector before CIAP treatment and comparing with non-filled-in vector DNA. Analyse the ligated DNA by agarose gel electrophoresis, or by *in vitro* packaging. The latter assay method gives a quantitative result.

7.3 Preparation of insert DNA

7.3.1 Complete digestion
The method for this is detailed in *Protocol 5*.

7.3.2 Partial digestion
The most convenient way to generate random DNA fragments for cloning into cosmid and phage λ vectors is to digest the DNA partially with a restriction enzyme recognizing a tetranucleotide sequence. The enzymes *Sau*3AI and *Mbo*I are commonly used because of their compatibility with *Bam*HI. The advantages of using these enzymes are as follows.

- Both recognize the sequence 5'-GATC-3', which occurs on average every 4^4 (256) bp in random sequence DNA. To generate 20 kb and 40 kb DNA fragments therefore, it is only necessary to cleave 1/80 and 1/160 of the available sites, respectively.

- Partial digests which are independent of DNA concentration and time can be reproducibly obtained by simultaneously adding *dam* methylase to the cleavage reaction (15). The two enzymes compete for binding to the target sites in the DNA. Sites that the restriction enzyme reaches first are cleaved, whereas sites that the methylase reaches first are modified, preventing subsequent cleavage.

- Self ligation of *Sau*3AI and *Mbo*I-cleaved DNA can be avoided by partial fill-in of the 4 bp overhang with dATP and dGTP. However, the 5'-GA ends are compatible with partially filled in *Xho*I or *Sal*I sites.

Details of the procedure used to generate partially digested and partially filled in DNA purified as described in *Protocol 6* is detailed in *Protocol 10*. Points to note are that in contrast to the published methods (15), the amount of *dam* methylase is kept constant and the amount of *Mbo*I or *Sau*3AI is decreased. This gives better results in our hands. The degree of partial digestion can be monitored in two ways:

(a) by Southern blotting: remove an aliquot of each partial digest reaction, separate on an agarose gel, blot, and probe with a probe specific for an interspersed repetitive sequence (e.g. BLUR 8 (16); *Figure 3*);

(b) by mixing an aliquot from the reaction with 100 ng of unmethylated λ DNA and incubating at 37°C. Because cleavage of the λ DNA parallels that of the YAC DNA (*Figure 3*), the need for Southern blotting is obviated.

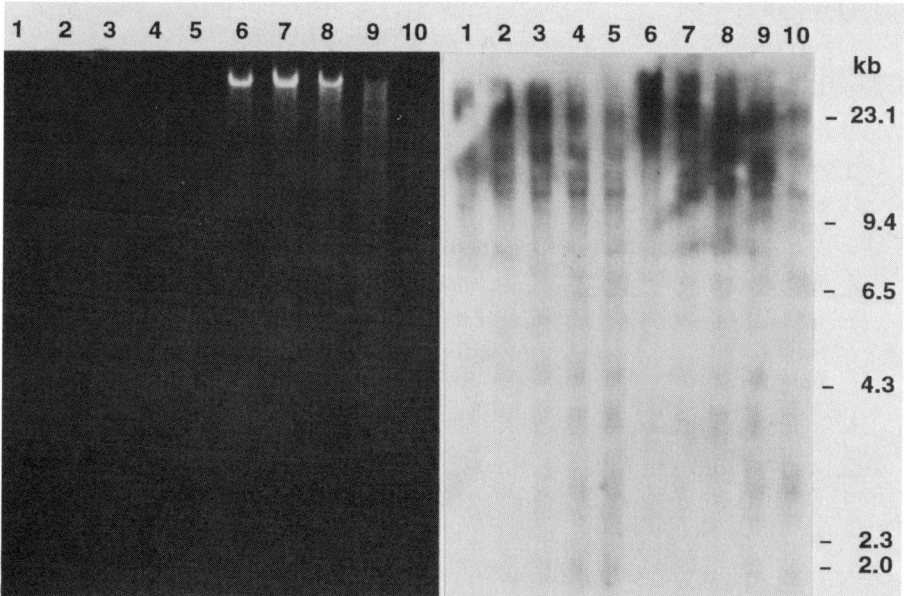

Figure 3. Analysis of partial digests of YAC DNA purified by preparative PFGE. Partial digests were performed on five 25 μl aliquots (~7 ng of YAC DNA per aliquot) as described in *Protocol 9*. Before incubating the reactions at 37°C for 60 min, 5 μl from each was removed, mixed with 100 ng of unmethylated λcl857 DNA, and incubated at 37°C for 60 min. 5 μl of DNA (~7 ng) from each of the five partial digest reactions (lanes 1–5), together with the aliquots containing the λ DNA (lanes 6–10), were subjected to electrophoresis on a 0.5% (w/v) agarose mini-gel (left panel), and alkali-blotted on to GeneScreen *Plus* (NEN Research Products) nylon membrane (right panel). After probing with the human repeat probe BLUR 8 (16), the filter was washed in 2 × SSC at 50°C before autoradiography for two days at −70°C. Size marker fragments are λ DNA digested with *Hind*III.

Protocol 10. Partial digestion by *dam* methylase competition

Equipment and reagents

- Agarase-treated DNA (*Protocol 6*)
- *Dam* methylase and S-adenosylmethionine (New England Biolabs)
- *Mbo*I or *Sau*3AI
- 10 × TAK buffer: 330 mM Tris-acetate pH 7.9, 650 mM potassium acetate, 100 mM magnesium acetate, 5 mM dithiothreitol
- 0.1 M EDTA pH 8.0
- Nylon membrane
- Phenol (*Protocol 7*)
- Chloroform:isoamyl alcohol (24:1)
- 100% ethanol
- 5 M ammonium acetate
- Dextran T-500 (Pharmacia): 10 mg/ml in water (sterilize by autoclaving)
- TE buffer (*Protocol 3*)
- Klenow fragment of DNA polymerase I
- 10 × fill-in buffer: 0.5 M Tris–HCl pH 7.5, 0.1 M MgCl$_2$, 1 mM DTT, 500 μg/ml BSA, 10 mM dATP, 10 mM dGTP
- Unmethylated λcI857 DNA (Promega)
- 1.5 ml microcentrifuge tubes
- Microcentrifuge
- Water baths or heating blocks at 37°C and 68°C
- Micropipettes and tips
- Mini-gel apparatus

Method

1. Prepare a reaction mixture containing 150 μl of purified YAC DNA, 30 μl of 10 × TAK buffer, 1 μl S-adenosylmethionine (32 mM), and 119 μl water.

2. Set up five microcentrifuge tubes. Dispense 100 μl into tube 1 and 50 μl into tubes 2–5.

3. Add 16 units of *dam* methylase to tube 1, and 8 units to tubes 2–5.

4. Add 0.5 units of *Mbo*I to tube 1, mix thoroughly and transfer 50 μl to tube 2. Mix well and continue the two-fold serial dilution through to tube 5. The *dam/Mbo*I ratio will vary from 32:1 in tube 1 to 512:1 in tube 5.

5. Incubate the tubes at 37°C for 60 min.

6. Add 10 μl 0.1 M EDTA to tubes 1–4, and 20 μl to tube 5. Incubate at 68°C for 10 min.

7. Remove a 5 μl aliquot, subject it to electrophoresis on a 0.5% (w/v) mini-gel, and transfer to nylon membrane. Hybridize the membrane with a ^{32}P-labelled DNA probe specific for an interspersed repetitive sequence (e.g. BLUR 8 (16) in the case of a YAC-containing human DNA insert, *Figure 3*) to check the size distribution of the partial digest fragments. Alternatively, remove 5 μl of each digest in step 4 and mix with 100 ng of unmethylated λ DNA and, after stopping the reaction (step 6), analyse the DNA fragment sizes on a 0.5% (w/v) agarose gel by ethidium bromide staining (*Figure 3*). Store the remainder of the partial digests at −20°C.

8. Pool the digests containing DNA fragments of the appropriate size,

and extract with an equal volume of phenol and chloroform:isoamyl alcohol. Add an equal volume of 5 M ammonium acetate, 2 μl of Dextran T-500, and 2 volumes of ethanol. Mix and incubate for 10 min at room temperature.

9. Pellet the precipitated DNA by centrifuging for 15 min in a microcentrifuge, dry the pellet briefly under vacuum, and resuspend in 17 μl TE buffer.

10. Add 2 μl 10 × fill-in buffer and 1 μl Klenow fragment. Incubate for 30 min at 37°C.

11. Extract with phenol and recover the DNA as described in step 8. Resuspend in 10 μl TE buffer.

7.4 Ligation

The most significant factor in the ligation of DNA is the concentration of compatible DNA termini in the reaction mixture. Depending on whether the cloning vector is a plasmid, or phage λ and cosmids, the optimal conditions for generating a significant yield of functional recombinant molecules are different. Although optimum values for the generation of concatemeric or circular DNA molecules can be estimated from theoretical considerations, by necessity such calculations assume that the exact concentrations of each DNA species in the ligation reaction are known and that all the DNA termini are undamaged. For PFGE-purified DNA, there will not usually be enough DNA to determine the optimal concentrations and the ratio of vector DNA to insert DNA. As a guide, suitable conditions are as follows.

(a) For a plasmid the size of pUC18, the ligation reaction should contain about 20 μg of vector DNA per ml, and an equimolar or greater concentration of DNA insert termini. The fraction of vector DNA converted to monomeric plasmid/foreign DNA chimeras increases rapidly to a maximum as foreign DNA concentration increases, and then decreases slowly (17).

(b) For phage λ or cosmid cloning, the ligation reaction should contain approximately 100 μg of vector DNA per ml. The amount of insert DNA will be variable, depending on the yield of gel-purified material, but a molar excess of dephosphorylated vector promotes a high cloning efficiency of small amounts of DNA, and reduces the probability of co-ligation of insert fragments.

7.5 Introduction into *E. coli*

Recombinant DNA molecules are introduced into *E. coli* either by transformation of competent cells with ligated DNA (plasmids), or infection of cells with phage λ particles containing *in vitro* packaged ligated DNA (phage λ or cosmids). Competent cells and *in vitro* packaging extracts can either be purchased commercially, or prepared in-house. Commercial products are very reliable and can give transformation and *in vitro* packaging efficiencies in excess of 10^9 transformants per μg of supercoiled plasmid DNA and wild-type phage λ DNA, respectively (e.g. Epicurian coli cells and Gigapack extracts from Stratagene). Although these products are many times more expensive than cells or extracts prepared in the lab, it is usually more economical for workers to use them, rather than to expend time and effort making their own. In cases where such large quantities of cells or extracts are used that it would be improvident to purchase them, home-made reagents may be worth the investment of time and effort. As a description of the preparation of competent cells and packaging extracts is beyond the scope of this chapter, the reader is referred to reference (17) for suitable protocols.

Plasmid DNA can also be introduced into *E. coli* by high voltage electroporation. Transformation efficiencies of up to 10^{10} transformants per μg of DNA have been obtained by optimizing various parameters. The preparation of cells for electroporation is easier than preparing competent cells (mid-log phase cells are washed with water and resuspended in 10% glycerol (18)), but it does require the purchase of an electroporator. Suitable units and ready prepared cells can be obtained commercially (e.g. the Electroporator and Electrocomp cells from Invitrogen Corporation).

The efficiency with which recombinant genomes are recovered as clones in *E. coli* is influenced considerably and selectively by host genetic background. The generation of *E. coli* strains with minimal abilities to restrict, recombine, and rearrange foreign DNA has been one of the most significant developments of recent years (9, 19). The main barriers to the establishment of cloned DNA in *E. coli* are

- the presence of restriction systems that cleave DNA lacking methyl groups at specific sites (*Eco*K, or *hsdR* system) or containing methyl groups at others (*Mcr* system);
- the recombination characteristics of the particular *E. coli* host.

For the propagation of plasmid and cosmid clones, use hosts that are deficient in all known restriction systems (*mcr*⁻, *hsdR*⁻) as well as being recombination deficient (*recA*⁻). Bacterial strains with this genotype are available commercially (e.g. XL1-Blue MR from Stratagene). Because host recombination systems are critical to the production of concatenated phage DNA (the substrate for *in vivo* packaging), consider the Rec genotype of the host in the light of its effect on the yield of phage and recovery of phage clones. *recD*⁻ strains which allow recombinant phage to obtain a large burst size as a consequence of rolling circle replication are now commonly used. A suitable strain is DL538 (*mcr*⁻, *hsdR*⁻, *recD*⁻ (20)) or NM722 (*NBL Gene Sciences Ltd*).

8. Clone analysis

8.1 Clones from genomic DNA

If an interspecies somatic cell hybrid is used as the source of plugs for preparative PFGE, the first step in the characterization of the resulting plasmid or phage clones will be the identification of those containing inserts from the target genome. In the case of a mouse hybrid containing a single human chromosome, achieve this by screening plasmid colony lifts, or phage plaque lifts with ^{32}P-labelled total human DNA, followed by washing of hybridized filters at moderate stringency (1 × SSC at 60°C). Because of the limited insert size of plasmid and phage λ insertion vectors (up to 10 kb), a small percentage of clones containing human DNA inserts will be missed since they are not big enough to contain a member of an interspersed repetitive sequence family.

Analyse cloned inserts by agarose gel electrophoresis after digesting isolated clone DNA with the appropriate restriction enzyme. At this stage, double-check the origin of the inserts by hybridizing a Southern blot of the digested DNA with the differential probe used to screen the initial library. The presence of more than one insert in a clone indicates either the co-ligation of non-contiguous insert fragments, or the incomplete digestion of the purified DNA before cloning. If clones such as these are encountered, hybridization of digested clone DNA with total DNA from the species used to construct the hybrid (e.g. mouse DNA in the case of a mouse–human somatic cell hybrid) may clarify the matter. If a clone contains a human and a mouse insert, then co-ligation artifacts are likely to be the cause of the multiple inserts. Clones containing very small inserts (≤ 1 kb) can also be discarded because they were identified by repetitive sequence hybridization and are therefore unlikely to contain adequate single-copy component for subsequent use as probes.

Identification and physical mapping of clones from the target DNA fragment involves the following steps:

• Hybridization of clones to Southern blots of DNAs from somatic cell hybrids or cell lines carrying deletions or translocations involving the target DNA fragment. Appropriate clones should hybridize only to hybrids containing the correct chromosome, and should show a pattern of hybridization to cell line DNAs consistent with the chromosomal rearrangement (e.g. if a deletion removes the region of the chromosome containing the target DNA fragment, a clone from that fragment should not hybridize to that DNA). In the case of clones from DNA fragments on the X chromosome, X dosage should be demonstrable between male and female DNA.

• Hybridization to Southern blots of PFGE gels of genomic DNA. Here, appropriate clones should show a pattern of hybridization to digests with infrequently cutting enzymes consistent with the long-range map for this region of DNA. Partial restriction digests can also be used with a hybridization pattern characteristic of a regional clone assignment.

It is important that both approaches be used when characterizing clones, as a misassignment at this stage can result in a lot of wasted time and effort later on. Examples of how these approaches have been used for the chromosomal localization and physical mapping of band-specific clones can be found in refs 3–5.

When using band-specific clones as start sites for chromosome walking, an important consideration is the number of steps required to achieve 'closure' (i.e. unbroken coverage of the entire region). This is dependent on the maximum gap between adjacent start sites (see *Figure 5* in ref. 3) and the size of the DNA fragments isolated at each step of the walk. As the size of gap can vary from as little as 10 kb to greater than 200 kb, 'closure' of large fragments is best achieved using YAC clones. YAC libraries can be screened by hybridization or PCR. If the band-specific clones have been generated using a vector that permits the rescue of single-stranded DNA for sequencing (e.g. pBluescript II or λZAP from Stratagene), then PCR screening will be the method of choice. Although YAC clones have the advantage of size for mapping purposes, they are less convenient as a source of DNA for the detailed analysis of a given genomic region, therefore the preparation of sublibraries may be the next step in the analytical programme.

8.2 Clones from purified YACs

Since all clones are derived from YAC DNA, characterization is much easier than for genomic band-specific clones (Section 8.1). At this stage, clones can either be screened directly for the presence of particular gene sequences, or ordered into a contig by repetitive sequence fingerprinting (21), Alu-PCR (22), or by hybridization of end-specific RNA probes (23). The first approach has been used for the construction of cosmid YAC subclone contigs, whereas the latter two have been used for the construction of phage λ contigs. An example of a phage contig derived from a PFGE-purified YAC, and ordered by hybridization of end-specific RNA probes, is shown in *Figure 4*.

Depending on the background level of non-recombinant clones in the clone collection, prior screening with ^{32}P-labelled DNA may be necessary at this stage. As the level of non-recombinants varies from batch to batch, and from vector to vector, it is probably worthwhile pre-screening the clones. As a guide, approximately 10% (10 of 121) of phage resulting from the subcloning of a 160 kb YAC from the human dystrophin gene were non-recombinant. It is virtually impossible to reduce background levels below this.

Although the number of clones needed to give a 99% probability of cover-

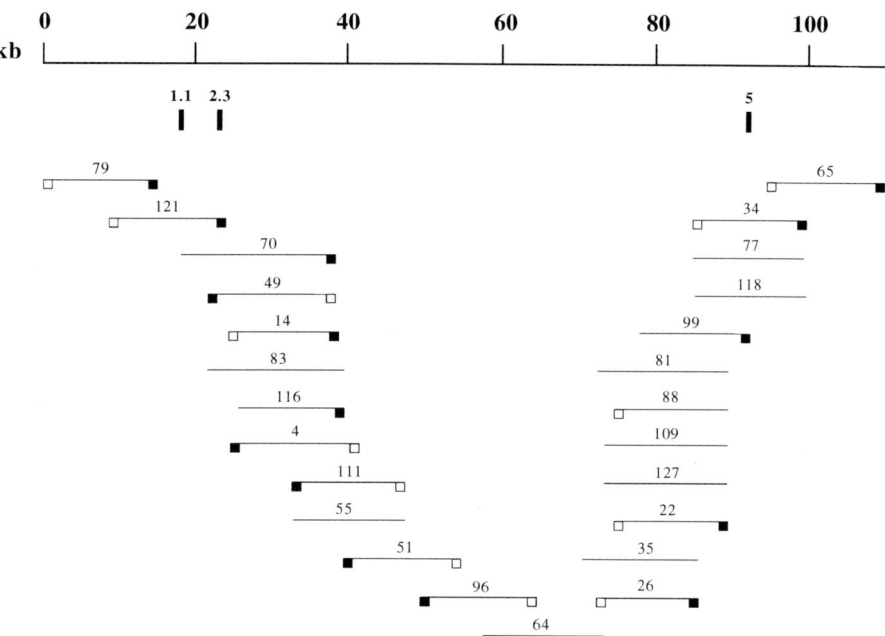

Figure 4. Contig of phage λ clones derived from a PFGE-purified YAC containing part of the human dystrophin gene. The YAC was purified and subcloned using vector λEMBL3*cos*W as described (14). Phage were ordered into an overlapping set by hybridization of end-specific RNA probes from individual clones back to plaque lifts of gridded phage subclones using the methodology in reference 23. Open and closed boxes indicate insert ends used to generate T7 and SP6 RNA probes, respectively, during contig construction. The three thick vertical bars above the contig indicate the positions of probes pXJ1.1 (26), pXJ2.3 (27), and clone 5, a 1.8 kb *Eco*RI fragment derived from an 840 kb genomic *Sfi*I fragment containing the 5′ region of the human dystrophin gene enriched by preparative PFGE (3).

ing a YAC calculated using the formula of Clark and Carbon (24) will be adequate for most purposes, ref. 25 suggests that 2–4 times this number may be needed to ensure the isolation of sequences that clone poorly. For example, clones containing the right arm of the YAC vector are frequently underrepresented in sublibraries (21, 23) compared with clones containing the left arm, because of differences in the number of *Mbo*I restriction sites between the right and left vector arms.

Acknowledgements

PW would like to acknowledge financial support from the Muscular Dystrophy Group of Great Britain and Northern Ireland, the Medical Research Council, the Nuffield Foundation, and Southampton University Research Fund.

References

1. Birren, B. and Lai, E. (1993). *Pulsed field gel electrophoresis: a practical guide.* Academic Press, San Diego.
2. Southern, E. M., Anand, R., Brown, W. R. A., and Fletcher, D. S. (1987). *Nucleic Acids Res.*, **15**, 5925.
3. Anand, R., Honeycombe, J. R., Whittaker, P. A., Elder, J. K., and Southern, E. M. (1988). *Genomics*, **3**, 177.
4. van de Pol, T. J. R., Cremers, F. P. M., Brohet, R. M., Wieringa, B., and Ropers, H.-H. (1990). *Nucleic Acids Res.*, **18**, 725.
5. Michiels, F., Burmeister, M., and Lehrach, H. (1987). *Science*, **236**, 1305.
6. Larin, Z., Monaco, A. P., and Lehrach, H. (1991). *Proc. Natl. Acad. Sci. USA*, **88**, 4123.
7. Couto, L. B., Spangler, E. A., and Rubin, E. M. (1989). *Nucleic Acids Res.*, **17**, 8010.
8. Whittaker, P. A. (1991). *Clin. Biotech.*, **3**, 67.
9. Murray, N. E. (1991). In *Methods in Enzymology* (ed. J. Miller), Vol. 204, pp. 280–301. Academic Press, New York.
10. Short, J. M., Fernandez, J. M., Sorge, J. A., and Huse, W. D. (1988). *Nucleic Acids Res.*, **16**, 7583.
11. Evans, G. A., Lewis, K., and Rothenberg, B. E. (1989). *Gene*, **79**, 9.
12. Zabarovsky, E. R. and Allikmets, R. L. (1986). *Gene*, **42**, 119.
13. Whittaker, P. A. and Wood, L. (1994). *Gene*, **138**, 227.
14. Whittaker, P. A., Mathrubutham, M., and Wood, L. (1993). *Trends Genet.*, **9**, 195.
15. Hoheisel, J. D., Nizetic, D., and Lehrach, H. (1989). *Nucleic Acids Res.*, **17**, 4571.
16. Jelinek, W. R., Toomey, T. P., Leinwand, L., Duncan, C. H., Biro, P. A., Choudary, P. V., *et al.* (1980). *Proc. Natl. Acad. Sci. USA*, **77**, 1398.
17. Sambrook, J., Fritsch, E. F., and Maniatis, T. (1989). *Molecular cloning, a laboratory manual* (2nd edn). Cold Spring Harbor Laboratory Press, New York.
18. Bothwell, A., Yancopoulos, G. D., and Alt, F. W. (1990). *Methods for cloning and analysis of eukaryotic genes.* Jones and Bartlett, Boston.
19. Kaiser, K., Murray, N. E., and Whittaker, P. A. (1995). In *DNA cloning* (2nd edn) (ed. D. M. Glover), Vol. 1, pp. 37–84. IRL Press at Oxford University Press, Oxford.
20. Whittaker, P. A., Campbell, A. J. B., Southern, E. M., and Murray, N. E. (1988). *Nucleic Acids Res.*, **16**, 6725.
21. Bellane-Chantelot, C., Barillot, E., Lacroix, B., le Paslier, D., and Cohen, D. (1991). *Nucleic Acids Res.*, **19**, 505.
22. Pieretti, M., Tonlorenzi, R., and Ballabio, A. (1991). *Nucleic Acids Res.*, **19**, 2795.
23. Whittaker, P. A., Wood, L., Mathrubutham, M., and Anand, R. (1993). *Genomics* **15**, 453.
24. Clarke, L. and Carbon, J. (1976). *Cell*, **9**, 91.
25. Zilsel, J., Ma, P. H., and Beatty, J. T. (1992). *Gene*, **120**, 89.
26. Ray, P. N., Belfall, B., Duff, C., Logan, C., Kean, V., Thompson, M. W., *et al.* (1985). *Nature*, **318**, 672.
27. Malhotra, S. B., Hart, K. A., Klamut, H. J., Thomas, N. S. T., Bodrug, S. E., Burghes, A. H. M., *et al.* (1988). *Science*, **242**, 755.

5

Yeast artificial chromosome cloning using PFGE

P. OUGEN and D. COHEN

1. Introduction

The major goal of the human genome project includes the isolation of the entire human genome in overlapping clones and the development of physical maps of the cloned DNA. Cloning into yeast artificial chromosomes (YAC) represents the method of choice for genome mapping analysis in the megabase range.

Since the first cloning in yeast (1), the availability of YAC libraries containing clones with large inserts from a variety of cells, ranging in size from 250 kb to 700 kb (2–6), provides a way to bridge the resolution gap between the level of linkage, or cytogenetic mapping, and the level of conventional recombinant technology. The development of pulsed field gel electrophoresis (PFGE) as an advanced technology used to prepare, analyse, and clone megabase molecules has had a large effect on the construction of YAC libraries.

Contour clamped homogeneous electric field (CHEF, 7–10) has a major advantage as a method for PFGE in that the electronic switching of fields can be achieved rapidly (5 sec) or slowly (90 sec to 15 min) for the separation of molecules smaller than 100 kb or larger than 1000 kb (and even up to 3 Mb) using a fixed angle at 120°. It provides excellent and reproducible separation with respect to the linearity of separation of molecules of increasing size. This chapter presents preparative and analytical migrations using CHEF-DRII (BioRad) and Pulsaphor (Pharmacia) apparatus.

Field inversion gel electrophoresis (FIGE, 11–12) which is a PFGE method used preferentially for analytical migration, presents a periodic inversion of electric field in order to achieve separation of large DNA molecules with a field interaction angle of 180°. This method requires a pulse for a longer time in the forward direction than in the reverse, the differential usually being a ratio of 3:1, forward to reverse. When the pulse times are progressively increased, while always maintaining the 3:1 ratio between forward and reverse pulse times, separations are achieved which are consistent with the molecular weight.

The cloning protocol developed to achieve the construction of a 'Mega-YAC' human library includes the embedding of intact human DNA in agarose (to prevent shear damage of high molecular weight DNA), partial digestion of agarose-embedded DNA, pulsed field gel electrophoresis for extraction and purification of very large digested and ligated DNA fragments (CHEF), analysis of the quality of transforming DNA (CHEF) and insert size (FIGE, CHEF), and modifications of classical yeast spheroplast preparations and transformation protocols (by incorporating polyamines in the cloning procedure).

2. Preparation of pYAC4 vector

The vector system which is used for the cloning of very large exogenous DNA fragments is the pYAC4 vector described by Burke *et al.* (1) in 1987. The pYAC4 vector contains an ampicillin resistance gene (*Amp*^{r}), a pBR ori (origin of replication) for replication in bacterial cells, two yeast selectable markers, and the cloned yeast genes encoding essential biosynthetic enzymes, *URA*3 (uracil), *TRP*1 (tryptophan), and *HIS*3 (histidine). The vector also contains the chromosomal sequences required for artificial chromosome maintenance, that is centromeric DNA (*CEN*4), autonomously replicating DNA (*ARS*), and telomeric DNA (*TEL*) which are derived from the ciliate eukaryote, *tetrahymena*. The vector contains a unique *Eco*RI cloning site, located in the intron of the *SUP*4 tRNA gene. The pYAC4 vector is replicated as circular molecules in *E. coli* and purifed by centrifugation in a CsCl gradient.

Before ligation to the large DNA molecules, the vector is prepared by digestion with *Bam*HI as outlined in *Protocol 1*. This removes the *HIS*3 fragment which is irrelevant to the cloning procedure but acts as a spacer fragment between the inverted *TEL* sequences. This digestion liberates the ends of the telomeres which are dephosphorylated to prevent recircularization of the vector and the *Bam*HI spacer fragment carrying the *HIS*3 gene will not ligate to anything. After phenol–chloroform extraction to eliminate the phosphatase enzyme, the unique cloning site is opened with *Eco*RI followed by a second phenol–chlorophorm extraction. In order to obtain the most efficient ligation of the vector to the cloning DNA, the cloning site is not dephosphorylated.

This procedure makes it possible to evaluate indirectly the ligation efficiency, since the three pYAC4 restriction fragments will ligate amongst themselves in a predictable way (right–right (8 kb), right–left or left–right (10 kb), left–left (12 kb)).

Protocol 1. Preparation of the pYAC4 vector

Equipment and reagents

- 10 × ligase buffer: 660 mM Tris–HCl, 50 mM MgCl$_2$, 10 mM DTT, 10 mM ATP, pH 7.5 (Boehringer Mannheim)
- 1% (w/v) agarose (SeaKem, FMC) in 1 × TBE (89 mM Tris-borate, 89 mM boric acid, 2 mM EDTA, pH 8.0)
- T4 DNA ligase

- *Eco*RI, *Bam*HI, and *Hind*III restriction enzymes and buffers
- Bacterial alkaline phosphatase (BAP) (Boehringer Mannheim)
- Small horizontal agarose gel electrophoresis apparatus
- Microcentrifuge

Method

1. Digest 0.5–1 μg of the pYAC4 plasmid with *Hind*III (1 unit/μg) for 2 h at 37°C. Analyse the products on a 1% (w/v) agarose gel to visualize four bands at 3.5 kb, 3.0 kb, 1.9 kb, and 1.4 kb (doublet). This procedure controls the integrity of the plasmid before preparing the pYAC4 arms for the cloning procedure.

2. Digest 0.5–1 mg of pYAC4 with *Bam*HI for 2 h at 37°C and check on a 1% (w/v) agarose gel for the presence of the 10 kb and 1.7 kb (*HIS*3 fragment) bands.

3. Add directly 0.05–0.1 unit of bacterial alkaline phosphatase (BAP) per μg plasmid DNA and incubate at 37°C for 1–2 h. Stop the reaction by heating at 75°C for 10 min. Extract with phenol–chloroform and precipitate it with ethanol.

4. Test the efficiency of the dephosphorylation of *Bam*HI vector ends by measuring the ability of these ends to ligate in the presence of T4 DNA ligase. To 10 μl of ligation reaction, add 1 μg of *Bam*HI-digested and BAP-treated plasmid DNA, 1 μl of T4 DNA ligase, and 1 μl of 10 × ligase buffer. Analyse on a gel and look for the presence of the 10 kb and 1.7 kb bands only.

5. Digest 0.5 to 1 μg pYAC4 plasmid with 1 unit/μg *Eco*RI for 2 h at 37°C. Use a gel to analyse the presence of 6 kb (left arm), 4 kb (right arm), and 1.7 kb (*HIS*3 fragment) bands.

6. Test the efficiency of the ability of the *Eco*RI extremities to ligate in the presence of the T4 DNA ligase. To 10 μl of ligation reaction, add 1 μg of *Eco*RI-digested plasmid DNA, 1 μl of T4 DNA ligase, and 1 μl of 10 × ligase buffer. Analyse with a gel and look for the presence of 12 kb, 10 kb, 8 kb, and 1.7 kb bands only.

---

3. Extraction and partial digestion of genomic DNA in agarose

3.1 Extraction of high molecular weight DNA in agarose

Very high molecular weight DNA fragments are impossible to prepare by conventional solution methods because they are extremely sensitive to shear damage. In order to prevent shear damage, protocols have been developed which allow DNA to be extracted from cells embedded in a matrix of low temperature melting restriction agarose as outlined in *Protocol 2*. Rare-cutting restriction enzymes or partial digestion are used to generate very large restriction fragments for YAC cloning.

Protocol 2. Extraction of high molecular weight DNA in agarose

Equipment and reagents

- Centrifuge (Megafuge from Heraeus)
- Phosphate-buffered saline (PBS): 10 mM KH$_2$PO$_4$, 15 mM NaCl, pH 7.4 (with K$_2$HPO$_4$)
- NDS: 100 mM EDTA, pH 8.0, 1% (w/v) N-lauroylsarcosine
- 1% (w/v) low melting temperature agarose (LMT; SeaPlaque, FMC) in PBS
- Perspex mould with slots of about 100 μl volume (10 × 6 × 1.5 mm)

- Proteinase K (Boehringer Mannheim)
- TE 10.5 (10 mM Tris–HCl pH 7.5, 5 mM EDTA)
- TE (10 mM Tris–HCl pH 7.5, 1 mM EDTA)
- Phenylmethylsulfonylfluoride (PMSF, 40 mg/ml stock solution in isopropanol, Sigma). (**Caution:** PMSF is toxic and should always be handled with gloves.)
- 50 mM EDTA pH 8.0

Method

1. Using cultured human cells, wash and pellet the cells in PBS by centrifugation (900 *g*, 5 min) at room temperature, count the number of cells and maintain the preparation at 37°C.

2. Mix the cells with 1% (w/v) LMT in PBS (or distilled water), gently homogenize the preparation and leave it at 37°C for 10 min. Pipette about 100 μl of the agarose mixture into the slots of a perspex mould, taped on one side, and leave the blocks to solidify at 4°C.

3. Incubate the plugs overnight in NDS buffer with Proteinase K (1 mg/ml final concentration) at 50°C, in order to lyse the cell membranes and remove all proteins and other molecules from the DNA.

4. Rinse plugs several times in TE 10.5

5. Incubate plugs in TE containing PMSF (40 μg/ml final) for 30 min at 4°C to inhibit all Proteinase K activity. Rinse twice in TE at room temperature to remove the PMSF and store at 4°C in 50 mM EDTA.

One million diploid cells should contain about 6 to 7 μg of DNA, assuming a normal genome-size of 6×10^9 bp of DNA. Agarose plugs each contain approximately 6 to 7×10^6 cells (40 μg of human DNA).

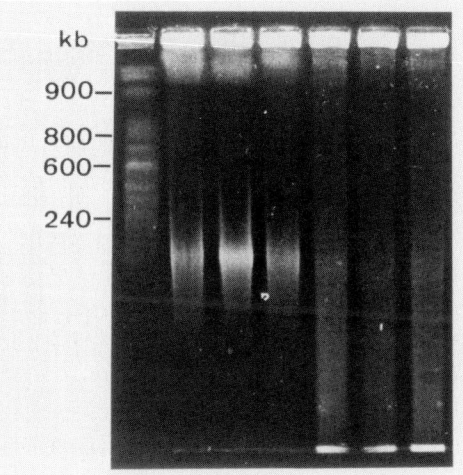

Figure 1. Analytical PFGE of DNA preparation: agarose plugs containing extracted human DNA were loaded on a 1% (w/v) agarose gel (SeaKem, FMC). The conditions of migration using a CHEF-DRII apparatus (BioRad) were 17 h at 190 V, in 0.5 × TBE buffer at 12°C with 90 sec pulses. After migration, the gel was stained with ethidium bromide. The left lane shows yeast chromosome markers from strain AB1380.

The quality of DNA extraction is assessed by analysing the amount of shear damage, which can be done using pulsed field gel electrophoresis (CHEF). In general, limited shearing is present in the unresolved zone (compression zone) and at the smaller sized DNA fragment (approximately at 10 kb) (*Figure 1*).

3.2 Generating large fragments of DNA by partial digestion

Large DNA fragments can be generated either by complete digestion with an enzyme whose recognition sequence occurs infrequently, such as *Not*I or *Mlu*I, or by partial digestion with an enzyme whose recognition sequence occurs frequently such as *Eco*RI, *Bam*HI, *Hind*III.

The construction of YAC libraries from partially digested DNA represents a good method to isolate overlapping YAC clones in order to establish contigs. To prepare partially digested, very high molecular weight DNA, different protocols for digestion may be used (see *Protocol 3*):

(a) limited concentration of cations (MgCl$_2$) (*Figure 2*);

(b) limited concentration of enzymes (*Eco*RI) (*Figure 3A, B*).

For all experiments, the plugs are washed several times in distilled water (to eliminate the EDTA) and then equilibrated in digestion buffer without enzymes and cations. It is commonly observed that a uniform partial

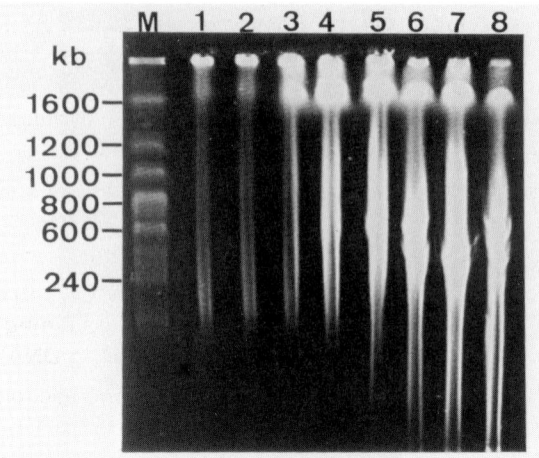

Figure 2. Analytical PFGE of partially digested DNA using various concentrations of $MgCl_2$. Plugs were loaded on a 1% agarose gel (SeaKem, FMC). The conditions of migration using a CHEF-DRII apparatus (BioRad) were 20 h at 190 V, in 0.5 × TBE buffer at 12°C with 90 sec pulses. After migration, the gel was stained with ethidium bromide. Lane M shows yeast chromosome markers from strain AB1380. Lane 1 shows the untreated DNA preparation. Lanes 2–8 shows the partially digested DNA preparations using various concentrations of $MgCl_2$ (0; 0.01; 0.05; 0.1; 0.5; 1; 10 mM) in the presence of *Eco*RI enzyme at 0.5 units/μg of DNA.

restriction digestion is very difficult to perform in agarose. In order to achieve homogeneous partial digestion of DNA in agarose, the plugs should be preincubated together with the enzyme to enhance their diffusion in the agarose matrix. Subsequently, when the $MgCl_2$ is added, the digestion activity is more homogeneous and efficient.

Protocol 3. Partial *Eco*RI restriction digestion

Equipment and reagents

- Distilled water
- *Eco*RI enzyme and *Eco*RI restriction enzyme buffer (100 mM NaCl, 10 mM Tris–HCl pH 8.0) without $MgCl_2$
- $MgCl_2$ (500 mM stock solution)
- TE 10.5 (*Protocol 2*)
- 50 mM EDTA pH 8.0

A. *Digestion using a limited concentration of $MgCl_2$*

1. Rinse the plugs in a large volume of distilled water (50 ml) and incubate them individually for 2 h at 4°C in 1.5 ml Eppendorf tubes with 100 μl of *Eco*RI restriction buffer containing only the *Eco*RI restriction enzyme (5 units/μg DNA) and no $MgCl_2$.

---

2. Add to separate tubes various concentrations (0.01 mM to 10 mM) of MgCl$_2$ and place them along with the tubes from step 1 at 37°C for 10 min. Start the enzymatic reaction by adding the pre-warmed MgCl$_2$ solutions to the tubes containing plugs, *Eco*RI enzyme, and *Eco*RI restriction buffer. Leave the digestion reaction for 1 h at 37°C.

3. Stop the reaction by adding a large volume (50 ml) of cold TE 10.5.

B. *Digestion using a limited concentration of enzyme*

1. Rinse the plugs in a large volume of distilled water (50 ml) and incubate them for 2 h at 4°C in *Eco*RI restriction buffer, containing the enzyme (*Eco*RI) at various concentrations (0.05 to 5 units/μg DNA).

2. Pre-warm MgCl$_2$ at 37°C for 10 min. To initiate the reaction, add MgCl$_2$ to 10 mM final concentration. Leave the digestion for 1 h at 37°C.

3. Stop the reaction by adding a large volume (50 ml) of cold TE 10.5 or EDTA 50 mM.

Protocol 3B, using a limited concentration of enzyme, is recommended for restriction enzymes such as *Eco*RI which have a star activity in these particular reaction conditions. But when *Eco*RI enzyme and MgCl$_2$ are added at the same time to start the restriction digestion, the partial digestion is less efficient than when MgCl$_2$ is added after pre-incubation with enzyme (*Figure 3B*).

In *Protocol 3*, test a sample of plugs from each DNA partial digestion by pulsed field gel electrophoresis in order to determine the optimal enzyme or cation concentration.

3.3 Analysis of the partial digestion

In order to determine the optimal conditions for generating very large DNA fragments, load the plugs containing the partially digested DNA in agarose gel slots and submit them to pulsed field gel electophoresis (CHEF-DRII, BioRad) under conditions that will separate DNA fragments sized 200 kb to 1600 kb (we use yeast chromosomal DNA in agarose blocks as size standards). After migration, stain the gel with ethidium bromide and assess three parameters (*Figure 2*, *Figure 3*):

(1) the concentration of DNA resting in the well: the concentration of partially digested DNA present in the well should decrease gradually as a function of the enzyme or MgCl$_2$ concentration compared to the well containing the untreated DNA;

(2) the concentration of DNA present in the unresolved zone: the DNA concentration present in the unresolved zone should increase gradually as a function of the enzyme or MgCl$_2$ concentration;

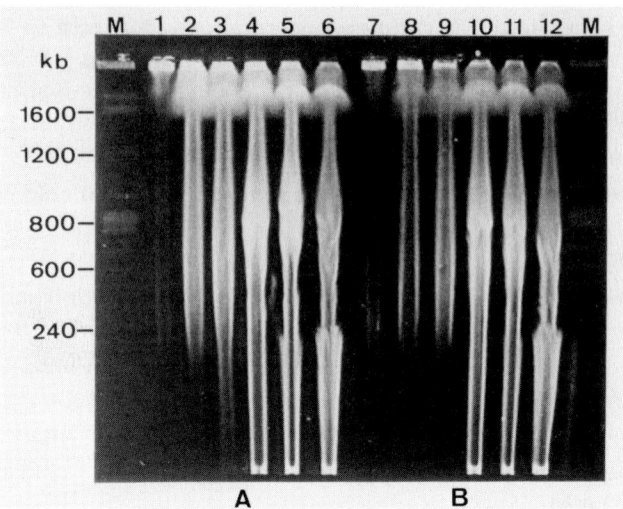

Figure 3. Analytical PFGE of partially digested DNA using various concentrations of *Eco*RI enzyme. Plugs were loaded on a 1% (w/v) agarose gel (SeaKem, FMC). The conditions of migration using a CHEF-DRII apparatus (BioRad) were 20 h at 190 V, in 0.5 × TBE buffer at 12 °C with 90 sec pulses. After migration, the gel was stained with ethidium bromide. Lane M shows the yeast chromosome markers from strain AB1380. Lanes 1 and 7 shows the untreated DNA preparation. Lanes 2–6 and 8–12 show partially digested DNA preparations using various concentrations of *Eco*RI enzyme (0.05; 0.1; 0.5; 1; 5 units/μg DNA) in the presence of 10 mM MgCl$_2$. Lanes 2 to 6 show plugs pre-incubated in the presence of *Eco*RI enzyme for 4 h at 4 °C then MgCl$_2$ added to start the restriction digestion. The plugs in lanes 8–12 had *Eco*RI enzyme and MgCl$_2$ added at the same time to start the reaction.

(3) the concentration of DNA present in the smear from the unresolved zone to the smallest DNA fragment: the DNA present in the smear represents all DNA fragments in a size range from the size determined by the pulse time to the smallest fragments removed from the partial digestion. This DNA concentration should increase as a function of the enzyme or MgCl$_2$ concentration.

The partial restriction digest protocols, using limited amounts of enzyme or cations, are reproducible and give an homogeneous partial digestion from the same extraction of DNA. The conditions of partial digestion should be determined for each preparation of DNA. After determination of the optimal concentration of enzyme or cation for the partial digestion, 10 to 20 blocks are treated using the same conditions.

4. Size separation of partially digested DNA

Since the first library constructed in our lab (5), we have increased the size selection of fractionated DNA fragments. The first library contained approxi-

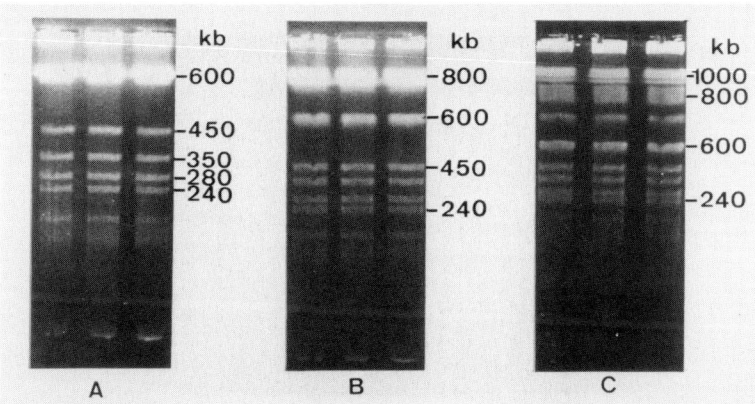

Figure 4. Analysis of the effect of pulse time on the size selection of large DNA fragments. Agarose plugs containing yeast chromosome markers from AB1380 were loaded on 1% (w/v) agarose gel (SeaKem, FMC). The conditions of migration using a CHEF-DRII apparatus (BioRad) were 18 h at 190 V, in 0.5 × TBE buffer at 12°C using various pulse times. After migration, the gel was stained with ethidium bromide. The pulse lengths were 35 sec (a), 50 sec (b), and 70 sec (c) to select 600 kb, 800 kb, and 1000 kb DNA fragments respectively.

mately 50 000 clones with an average size of 450 kb. This sizing was obtained using CHEF-DRII apparatus (BioRad) using migration conditions of 15 sec pulse time at 190 V, for 18 to 20 h at 12°C in 0.5 × TBE. These standard conditions lead to the selection of DNA fragments of 300 kb to 350 kb in the compression zone.

In order to increase the size of the selected DNA fragments, the pulse time was gradually increased while keeping the same voltage, run time, temperature, buffer, and concentration of agarose. The effect of a gradual increase in pulse time on the separation of yeast chromosomes from *Saccharomyces cerevisae* AB 1380 (240 kb to 2200 kb) is shown in *Figure 4*. For example, 35 sec pulses separate DNA fragments up to 600 kb, 50 sec pulses separate DNA fragments up to 800 kb, 70 sec pulses separate DNA fragments up to 1000 kb.

After partial digestion, large fragments of DNA are size fractionated by using conditions optimal for the desired size. *Figure 5* shows a size fractionation of partially digested DNA using 50 sec and 70 sec pulses to select 800 kb and 1000 kb respectively. The lanes containing yeast chromosome markers and a part of the DNA are cut from the gel and stained with ethidium bromide. The compression zone is evident and can be excised from the unstained part of the gel. The agarose matrix which contains partially digested and size-selected DNA fragments can be preserved in EDTA (50 mM) at 4°C or washed several times in distilled water to remove TBE and equilibrated in ligation buffer without any ligase or ATP, before ligation to the vector.

(a)

kb
—800
—600
—240

(b)

kb
—1000
— 800
— 600

Figure 5. First preparative CHEF migration. After partial digestion using *Eco*RI enzyme, the digested products are size-selected using a CHEF-DRII apparatus (BioRad). The 1% (w/v) LMT agarose gel (SeaPlaque, FMC) is subjected to electrophoresis at 190 V, using 50 sec pulses (a) to select 800 kb and 70 sec pulses (b) to select 1000 kb for 18 h at 12°C, in 0.5 × TBE buffer. The yeast chromosome markers are included on both sides of the gel to guide the recovery of the DNA products in the unresolved zone. The lanes with the yeast chromosome markers and a small part of the gel containing DNA are cut from the gel and stained with ethidium bromide. The unresolved zone (containing partially digested and size selected DNA fragments) could be determined. The corresponding part of the unstained gel is excised from the gel by aligning with the marker lane adjacent to the gel.

5. Ligation of DNA to YAC vectors in agarose

The ultimate goal of the ligation reaction is to insert only one fragment of exogenous DNA into the arm of the vector. In order to achieve this, the large DNA fragments are ligated to the YAC vectors in the presence of a large excess of vector as outlined in *Protocol 4*. Vector is added to the insert

DNA in a molar excess of 40:1, when the large molecules of DNA have been size selected at 350 kb, and 80:1 when the large molecules of DNA have been size selected at 800 to 1000 kb (the ratio increases gradually in proportion with the augmentation of the pulse time used for the size selection). This maintains an equal ratio of mass of vector to mass of DNA insert. In order to achieve the greatest ligation efficiency, do not dephosphorylate the cloning site. To achieve a homogeneous mixture of vector and insert, briefly melt the matrix of agarose which contains the DNA insert in the presence of the vector DNA and prior to the ligation reaction. In order to minimize the co-ligation events, dilute the preparation two-fold in the ligase buffer.

Several groups (6, 14) have added NaCl and polyamines to the ligation buffer (spermine and spermidine) in order to stabilize and protect the large DNA fragments during the melting and the ligation reaction. NaCl is an inhibitor of the ligase enzyme, and polyamines are capable of precipitating large DNA molecules. These effects may decrease the ligation efficiency.

Protocol 4. Ligation in agarose

1. Rinse 1 ml of agarose, recovered from the size fractionation gel, several times at room temperature in distilled water and in 1 × ligation buffer (see *Protocol 1*).

2. Place the solid agarose matrix in a tube. Add the vector DNA to the agarose matrix in a quantity equal to the estimated weight of insert DNA and melt the mixture in a 66°C water bath for 10 min and transfer to a 37°C water bath.

Either

3A. Add to the preparation 1 ml of 1 × ligation buffer with 20 units T4 DNA ligase (Amersham) preheated to 37°C, and mix gently. The ligation reaction is incubated at 37°C for 2 h, then transferred to room temperature overnight. Before the second size-fractionation migration, rapidly melt the agarose at 68°C and load it on to the electrophoresis gel, using a cut-off tip.

or

3B. Add 1 ml of 1 × ligation buffer without ligase to the mixture and load the preparation into a mould to reform it into a normal slice of agarose. Incubate the agarose slice (2 ml) in 10 ml of 1 × ligation buffer containing 20 units T4 DNA ligase at room temperature for 1 to 2 days. Load the slice of agarose into an electrophoresis gel slot (without melting), before the second size fractionation by PFGE.

This last step (3B) of *Protocol 4*, described by Chartier *et al.* (13) and modified in our lab, eliminates the second melting step and reduces the amount of DNA shearing.

6. Size separation of ligated DNA

Before transformation in yeast, the ligation products should be purified from the excess DNA vector using a second size-fractionation migration performed under the same conditions used for the first size fractionation. Place the agarose matrix containing the ligated DNA in the well of an electrophoresis gel (1% (w/v) LMT in 0.5 × TBE) and subject it to CHEF electrophoresis. After the migration, lanes containing the yeast chromosome markers and a part of the size-selected DNA are cut from the gel and stained with ethidium bromide. The unresolved zone is evident and can be excised from an unstained part of the gel. As expected, some very high molecular weight ligated DNA is recovered at the predicted size according to the pulse time used for the electrophoresis. Partially digested DNA molecules ligated to the vector are size-selected by the CHEF as before, and are recovered from the corresponding places (*Figure 6*).

Excess vector is eliminated during the second CHEF migration. As described previously in Section 2, the cloning site is not dephosphorylated, in order to increase the ligation efficiency and to give an indirect control of the ligation efficiency. This control is obtained by observing re-ligated vector-arm products after the second CHEF migration. If the ligation is complete, three lower bands appear in the gel, representing the ligation of the left–left arms

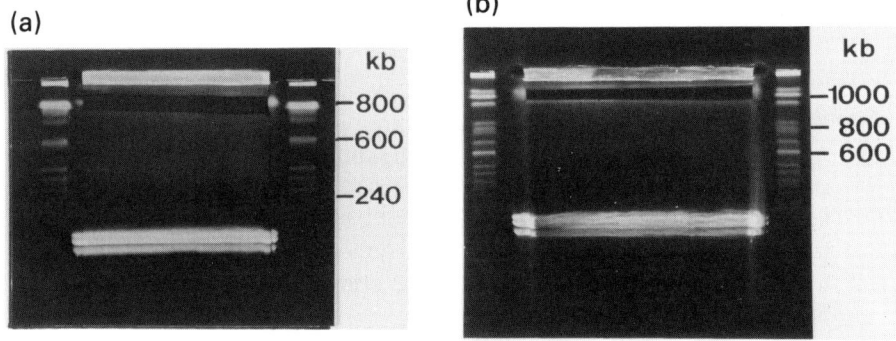

Figure 6. Second preparative CHEF migration. After ligation of the vector to the partially digested human DNA, the ligation products are size-selected using a CHEF-DRII apparatus (BioRad). The yeast chromosome markers are included on both sides of the gel to guide the recovery of the DNA products in the unresolved zone. The 1% (w/v) LMT agarose gel (SeaPlaque, FMC) is subjected to electrophoresis at 190 V, using 50 sec pulses (a) to select 800 kb and 70 sec pulses (b) to select 1000 kb for 18 h at 12°C, in 0.5 × TBE buffer. The lanes with the yeast chromosome markers and a small part of the gel containing DNA are cut from the gel and stained with ethidium bromide. The unresolved zone (containing digested and size selected DNA fragments) could be determined. The corresponding part of the unstained gel is excised from the gel by aligning with the marker lane adjacent to the gel. The lower bands represents ligation products of the two vector arms, an indirect indication of the ligation efficiency (left/left 12 kb; right/left and left/right 10 kb; right/right 8 kb).

(12 kb), the left–right and the right–left arms (10 kb), and the right–right arms (8 kb). The presence of two lowest bands at 6 kb (left arm) and 4 kb (right arm) suggests an incomplete ligation of vector arms and indirectly suggests an incomplete ligation of the partially digested DNA to the vector DNA.

7. Extraction and analysis of transforming DNA from agarose

7.1 Extraction of DNA from agarose

The ligated DNA is liberated from the agarose matrix before transformation into yeast as outlined in *Protocol 5*. It is common to melt the agarose matrix at 66°C for 10 min (2, 3) while some prefer to digest the agarose with agarase treatment before adding the transforming DNA to a spheroplast preparation (5, 6).

The gradual increase of the pulse time during preparative electrophoresis selects very large DNA molecules. In order to stabilize and protect large transforming DNA molecules during melting, agarase treatment, and transformation, the slice of agarose is first equilibrated in a specific 'CaS-buffer' (Ca: Calcium; S: Sorbitol) which contains

(a) NaCl, which is necessary for the agarase activity and enhances DNA stability;

(b) sorbitol to preserve the viability of the spheroplasts;

(c) polyamines (spermidine, spermine) to protect and stabilize the large DNA molecules from shearing during melting, agarase treatment, and transformation;

(d) $CaCl_2$ which conditions the DNA molecules at the same concentration in STC and PEG buffers, increases the permeability of the cell membranes, and possibly helps to increase DNA penetration.

Protocol 5. Agarase treatment

Equipment and reagents

- Distilled water
- CaS buffer: 30 mM NaCl, 0.7 mM spermidine, 0.3 mM spermine, 1 M sorbitol, 10 mM $CaCl_2$
- Water bath at 66°C
- Agarase (Gelase from Epicenter or Beta-agarase I from New England Biolabs)

Method

1. Wash a slice of agarose several times in distilled water to remove the TBE buffer and equilibrate it in CaS buffer.

Protocol 5. *Continued*

2. Transfer the agarose into a 10 ml tube and incubate for 5 min in a 66°C water bath.

3. Transfer the tube to 37°C and add agarase at 1 unit per mg of LMP agarose. Incubate the reaction for 1–2 h at 37°C.

4. Leave the tube at 4°C to verify the agarase efficiency, before transformation.

7.2 Analysis of transforming DNA liberated from agarose

Before transformation of the DNA into spheroplasts, analyse the quality of transforming DNA by PFGE. Load a sample of the agarose matrix containing ligated, size-fractionated DNA, washed in CaS buffer and digested with agarase, on to an electrophoresis gel. Run the gel for analytical migration using the same PFGE conditions as was done in the selection of the appropriate size in the preparative migration. In *Figure 7*, the DNA has been submitted to 90 sec pulses. After migration, transfer the gel to a nylon membrane. An

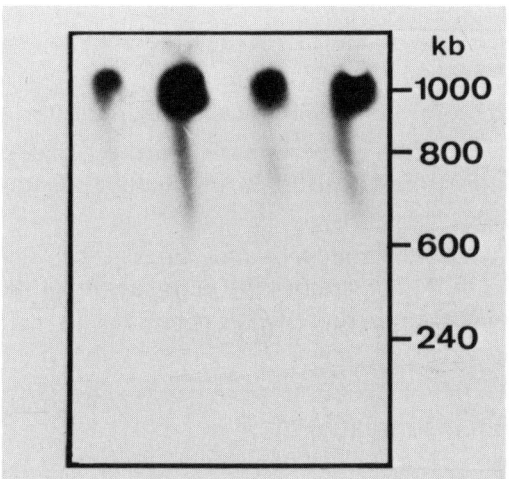

Figure 7. Analysis of transforming DNA using PFGE. Before transformation into the spheroplasts, a sample of the digested agarose containing ligated, size-fractionated DNA is submitted to an analytical electrophoresis using a CHEF-DRII apparatus (BioRad). The 1% (w/v) agarose gel (SeaKem, FMC) is subjected to electrophoresis at 190 V, using 90 sec pulses, for 20 h at 12°C, in 0.5 × TBE buffer, then transferred to a nylon membrane. An autoradiograph of the membrane hybridized with total human DNA is shown. The majority of the ligated products are present at 1000 kb (the size corresponds to the selected size used for the preparative CHEF electrophoresis present in *Figures 5b* and *6b*) with a small smear present at 600 kb.

autoradiograph of the membrane hybridized with a total human DNA probe shows that the majority of the ligated DNA molecules are present at 1000 kb (this size corresponds to the size selection, see *Figure 6b*) while a small smear extends to 600 kb. This treatment suggests that the CaS buffer is effective at stabilizing and protecting the very large DNA molecules during the melting and agarase treatment, before transformation.

8. Transformation of yeast spheroplasts

The efficiency of the transformation is limited by the size of the transforming DNA and the competence of the yeast. The transformation efficiency decreases gradually as the transforming DNA increases in size. In order to achieve a high transformation efficiency using very large DNA molecules, a modified version of the most common protocol described by Burger and Percival (15) is used. *Protocol 6* outlines the preparation of spheroplasts and *Protocol 7* describes the transformation of the spheroplasts with high molecular weight DNA. Different enzymes can be used to generate spheroplasts (Zymolyase from Seikagaku Kogyo, lyticase from Sigma, novozyme from Novobiolabs), and each have specific individual kinetics that must be determined in order to obtain the greatest transformation efficiency. To control the spheroplast procedure, prepare a range of various concentrations of enzyme (Zymolyase) prior to each transformation to be tested in parallel. Only one preparation of spheroplasts is used for the transformation.

Protocol 6. Spheroplast preparation

Equipment and reagents

- YPD solution: 1% (w/v) yeast extract (Difco), 2% (w/v) bactopeptone (Difco), 2% (w/v) D-glucose (Sigma); autoclave, adjust pH to 5.8. Add 1.5–2% (w/v) bactoagar (Difco) to prepare YPD agar plates
- 1 M sorbitol (Sigma)
- SCEM solution: 1 M sorbitol, 10 mM EDTA, 0.1 M Na-citrate pH 5.8; autoclave and add 30 mM β-mercaptoethanol before use
- STC buffer: 1 M sorbitol, 10 mM Tris–HCl pH 8, 10 mM CaCl$_2$; autoclave

- Zymolyase 20-T (20 000 units/g, Seikagaku Kogyo) stock solution: 50 mg/ml in 10 mM Tris pH 7.5, 1 mM EDTA, 50% (v/v) glycerol
- Haemocytometer
- Spectrophotometer
- *S. cerevisiae* strain AB1380 streaked fresh on to a YPD plate and grown for several days at 30°C
- Shaking incubator
- Standard light microscope with phase contrast
- Glass microscope slides and cover slips

Method

1. Inoculate a single colony of AB1380 from a YPD agar plate into 200 ml of YPD and grow at 30°C with vigorous shaking (200 r.p.m.), overnight, to give a culture density of 1–5 × 10^7 cells/ml (mid-log phase). Pellet the cells by centrifugation (900 *g*, 5 min) at 20°C.

109

Protocol 6. *Continued*

2. Resuspend the pellet and rinse cells in 1 M sorbitol. Calculate the number of the cells using a haemocytometer and make aliquots of 1.5 × 10⁹ cells resuspended in 5 ml SCEM.

3. Add Zymolyase-20 T at various concentrations (1–10 units/sample) and incubate the cells at 30 °C for 15 min with gentle agitation (80 r.p.m.).

4. Harvest the cells by centrifugation at 450 g at 20 °C for 5 min. Resuspend the pellet in YPD containing 1 M sorbitol and incubate at 30 °C for 30–60 min with gentle agitation.

5. Harvest the cells by centrifugation (450 g, 5 min) at 20 °C. Resuspend pellet in 5 ml STC buffer containing polyamines (0.7 mM spermidine and 0.3 mM spermine). The cells are now stable and ready for transformation.

6. Check the spheroplast preparations using

   (a) a spectrophotometer at $OD_{600\ nm}$; diluting each sample to 1/10 in distilled water and 1 M sorbitol. Estimate the OD of untreated cells (C), the OD of spheroplasts in water (W), and the OD of spheroplasts in sorbitol (S). The ratio of S/C should be close to 1; this value gives a good estimation of the quality of the spheroplast preparation. The optimal ratio W/S is approximately 0.5 (this is an average value calculated from all the preparations performed in our lab);

   (b) a microscope to compare the viability of cells in water and in sorbitol. 98% of cells are viable in sorbitol but are rapidly lysed in water to 100% lysis.

Protocol 7. Transformation of spheroplasts

Equipment and reagents

- PEG: 20% (w/v) polyethylene glycol (PEG 8000, Sigma), 10 mM Tris–HCl pH 8.0, 10 mM CaCl₂; filter and sterilize
- SOS: 1 M sorbitol, 6.5 mM CaCl₂, 0.25% (w/v) yeast extract (Difco), 0.5% (w/v) bactopeptone, 20 µg/ml uracil and tryptophan; filter and sterilize, pH 5.8
- TOP: 1 M sorbitol, 0.67% (w/v) yeast nitrogen base without amino acids (Difco), 1.5–2% (w/v) bactoagar (Difco); autoclave, pH 5.8, and add amino acids as required
- SORB: 0.9 M sorbitol, 3% (w/v) D-glucose, 0.67% (w/v) yeast nitrogen base without amino acids (Difco), 1.5–2% (w/v) bactoagar (Difco); autoclave, pH 5.8, and add amino acids as required
- Centrifuge
- SD: 2% (w/v) D-glucose, 0.67% (w/v) yeast nitrogen base without amino acids; autoclave, pH 5.8, and add amino acids as required
- amino acids (Sigma):

	Ura-	Ura-, Try-
Ade A 9795	1	1
Arg A 3909	4	—
His H 9511	2	2
Iso I 7383	6	6
Leu L 1512	6	6
Lys L 1262	5	5
Met M 2893	2	2
Phe P 5030	5	5
Thr T 1645	20	20
Trp T 0271	4	—
Tyr T 1020	5	5
Total	60	52
Used at	600 mg/l	520 mg/l

110

Method

1. After agarase treatment, add 10 to 20 µl samples of the ligated size-selected DNA to the spheroplast preparations (1.5×10^7 cells in 100 µl per transparent 10 ml tube).

2. Mix the cell suspension gently and incubate tubes at room temperature for 10 min.

3. Add 1 ml PEG buffer containing polyamines (0.7 mM spermidine and 0.3 mM spermine), mix gently and incubate the tubes for 10 min at room temperature.

4. Centrifuge tubes at 450 *g* at 20°C for 5 min, discard the supernatant, resuspend the pellet in 150 µl of SOS, and incubate at 30°C for 30–45 min.

5. Add 3 ml of TOP agar without uracil, preheated to 45°C, to each tube and transfer the contents to Petri dishes containing SORB agar without uracil.

6. Check for the presence of transformants (red colonies) after 3 to 4 days of incubation at 30°C.

7. Pick red colonies and transfer on to double selective plates (96-well microtitre) lacking uracil and tryptophan. Incubate them overnight at 30°C.

It is recommended that the primary transformants are transferred to SORB medium lacking only uracil. Clones that are growing in the first selective medium have a red phenotype characteristic of the interrupted *SUP* 4 tRNA gene, and are transferred to SD medium lacking uracil and tryptophan to verify the quality of YAC transformants and to control the presence of left and right vector arms. Analyse randomly chosen clones from each transformation by PFGE and hybridization with total human DNA or vector DNA probe.

9. Analysis of size of YAC clones using PFGE

9.1 Extraction of yeast DNA in agarose

Before analysing the size of cloned YAC inserts using PFGE, the high molecular weight yeast DNA molecules are extracted in agarose matrix as outlined in *Protocol 8*. This protocol is used to purify yeast chromosomes in agarose. The purified DNA is ready for PFGE analysis and restriction enzyme digestion.

Protocol 8. Extraction of high molecular weight yeast DNA for
          PFGE

Equipment and reagents

- YPD medium (*Protocol 6*)
- Centrifuge
- Incubating shaker at 30°C
- Distilled water
- 0.5% (w/v) LMT agarose in distilled water
- SCEM (*Protocol 6*)
- Zymolyase 20-T (*Protocol 6*)

- Perspex mould with slots of about 100 μl volume (10 × 6 × 1.5 mm)
- NDS buffer (*Protocol 2*)
- Proteinase K (*Protocol 2*)
- TE 10.5 (*Protocol 2*)
- 0.5 × TBE (*Protocol 1*)

Method

1. Culture a part of a red colony in 1–2 ml of YPD medium overnight at 30°C with strong shaking.

2. Recover yeast cells by centrifugation at 900 *g* at 20°C for 5 min and rinse the pellet in distilled water.

3. Resuspend yeast cells in 0.5% (w/v) LMT agarose dissolved in distilled water, at 3×10^8 cells/ml (2–3 μg of yeast DNA/100 μl) and load the preparation into the slots of the Perspex mould (100 μl) on ice.

4. Transfer each block into 1 ml of SCEM with 10 units/ml of Zymolyase 20-T, and incubate at 37°C for 2–3 h.

5. Remove the SCEM buffer and add an equal volume of NDS buffer with 1 mg/ml Proteinase K and incubate overnight at 50°C.

6. Rinse the block several times with TE 10.5 or 0.5 × TBE buffer.

7. For restriction enzyme digestion, treat plugs with PMSF (40 μg/ml final concentration, *Protocol 2*) to eliminate the Proteinase K, wash them in distilled water and then equilibrate them in restriction digestion buffer.

9.2 Analysis of insert size using PFGE

9.2.1 Analysis using FIGE

After each transformation, DNA from 10 to 20 randomly chosen YAC clones are analysed for molecular size by FIGE. This method requires a pulse for a longer time in the forward direction than in the reverse (see Chapter 1). The differential is usually a ratio of 3:1, forward to reverse. The conditions of migration are further explained in the legends of *Figures 8* and *9*.

Figure 8 shows the mean size range of random YACs from the last generation of the CEPH YAC library. Under specific conditions of pulse time (forward 6 sec to reverse 90 sec), the mean size is more than 900 kb. *Figure 9* shows the effect of the absence (A) or the presence (B) of polyamines during the transformation procedure. When the preparation of transforming

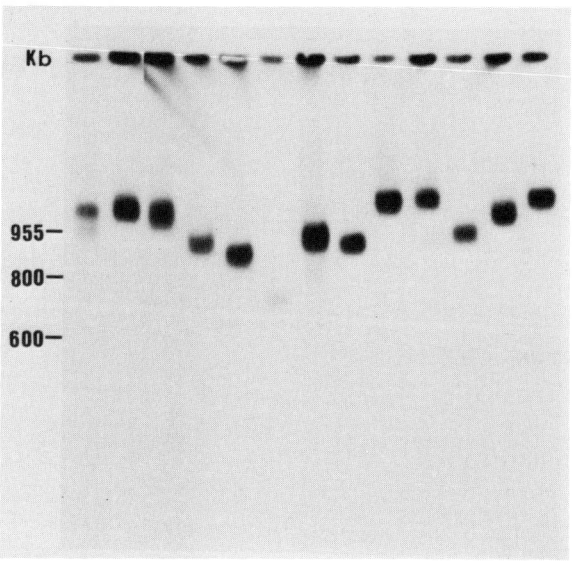

Figure 8. Analysis of the size of YAC insert DNA using FIGE. Agarose plugs containing YAC extracted DNA (from the last generation of YACs) were loaded and run on 1% (w/v) agarose gel (SeaKem, FMC) in 0.5 × TBE at 220 V for 22–24 h at 12°C using 6–90 sec pulses with a ratio of 3:1. A blot of this gel was hybridized to a total human DNA probe labelled with ^{32}P-dCTP. The majority of the YAC clones migrated at the size corresponding to the selected-fractionation size used for the preparative CHEF electrophoresis (*Figures 5b* and *6b*).

DNA and the transformation are performed in the absence of polyamines, the size of the YAC clones is heterogeneous and does not correspond to the size selected during the two CHEF preparative migrations. By contrast, the presence of polyamines during the preparation of transforming DNA and the transformation contribute to stabilization of the YAC size at the selected preparative size. However, as described by others (6, 14), the presence of polyamines during the preparation of transforming DNA and the transformation reduces 2–5 fold the number of transformants (data not shown).

9.2.2 Analysis using CHEF electrophoresis

In order to verify the YAC size observed by FIGE, DNA clones from the last generation of the CEPH library were analysed by CHEF using a Pulsaphor apparatus (Pharmacia). Larger stretched molecules require longer times for reorientation in a changed electric field, so in order to achieve a good separation of very large molecules (up to 2 Mb), a diminution of the voltage and an augmentation of the run time is required. The specific conditions are explained in detail in figure legends (*Figures 10* and *11*). The analysis of agarose plugs each containing DNA of individual clones and agarose plugs each containing DNA from 768 YAC clones (8 × 96) are

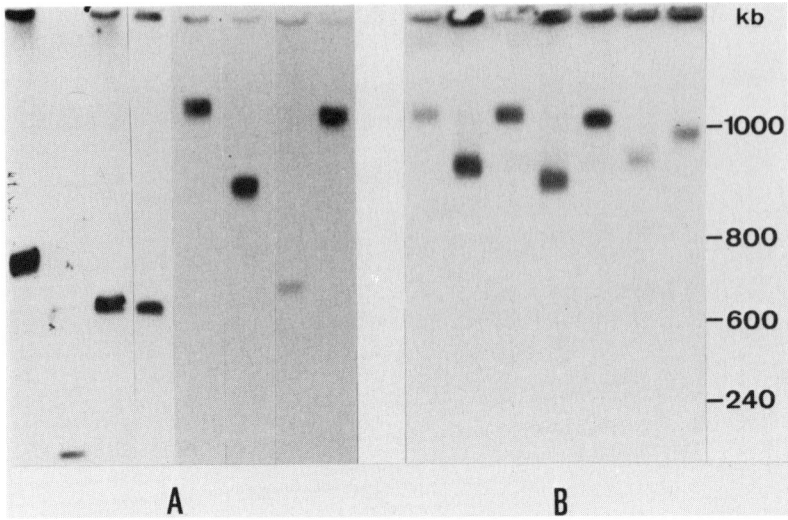

Figure 9. Analysis of the effect of polyamines on YAC insert size using FIGE. Agarose plugs containing YAC extracted DNA from clones obtained without (A) and with (B) polyamines during transformation were loaded and run on 1% (w/v) agarose gel (SeaKem, FMC) in 0.5 × TBE at 220 V for 24 h at 12°C using 6–90 sec pulses with a ratio of 3:1. A blot from this gel was hybridized to a total human DNA probe labelled with ^{32}P-dCTP. In the absence of polyamines (A) the YAC insert size appeared heterogeneous. In the presence of polyamines (B) the YAC clones were found at the size corresponding at the selective-fractionation size used for the preparative CHEF electrophoresis (*Figures 5b* and *6b*).

presented in *Figure 10* and *Figure 11*, respectively; *a* shows the ethidium bromide-stained gel and *b* shows the corresponding filter transfer probed with ^{32}P-dCTP-labelled pBR322.

After hybridization to the membrane, the size obtained from single DNA clones determined by CHEF migration are in agreement with the size determined by FIGE migration (*Figure 8*). The size range of clones is close to 1200 kb. This result emphasizes the high resolution of the FIGE system.

Electrophoresis of a pool of 96 YACs by CHEF contributes to the analysis of the size distribution of a part or a total library. *Figure 11* shows that the great majority of clones (from the last generation of the CEPH YAC library) are present from 800 kb to 1600 kb. Lane (i) shows a clone at 2850 kb size. Lanes 0 and 8 (*Figure 10*) and lane M (*Figure 11*) show chromosome markers from *S. pombe* (3.5 Mb chromosome is separated). Lane 9 and lane (a) show a mouse genomic YAC (3.3 Mb) donated by R. Rothstein.

A total YAC library of total human genomic DNA was constructed in increments with fragments ranging in mean size from 430 kb to 1200 kb (*Figure 12*) (16). The last generation of the YAC library called 'Mega-YAC'

(a)
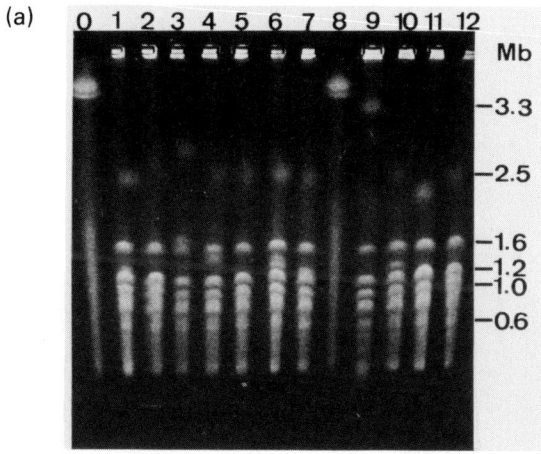

(b)
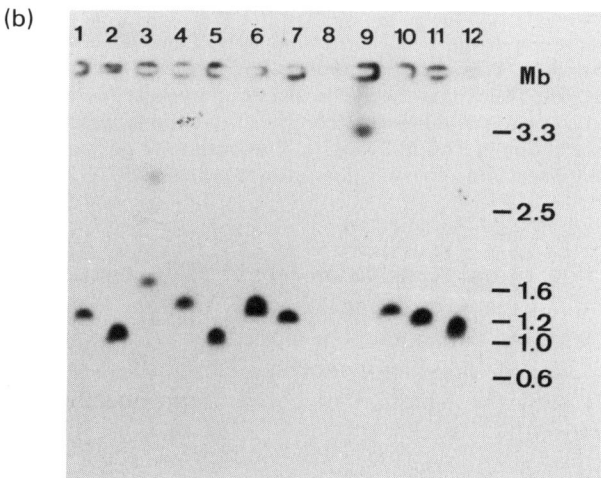

Figure 10. Ten randomly chosen clones were analysed by the CHEF 'Pulsaphor' apparatus (Pharmacia). The 1% (w/v) agarose gel (SeaKem, FMC) in 0.5 × TBE was subjected to electrophoresis at 70 V, using a 15 min pulse time, for 117 h at 12°C. All clones were size-distributed between 0.8 and 1.6 Mb. Lane 8 shows chromosome markers from *S. pombe*. Lane 9 shows a mouse genomic YAC (3.3 Mb). (a) Photograph of gel stained with ethidium bromide; (b) the corresponding transfer filter probed with ^{32}P-dCTP labelled pBR322.

(MARK IV and MARK V) contained 10 400 clones and 8000 clones with a mean insert size of 980 kb and 1.2 Mb, which were obtained when all crucial modifications were incorporated into the classical YAC cloning protocol. In spite of these changes, it always takes about two weeks to fully analyse the results and to determine the efficiency of any transformation.

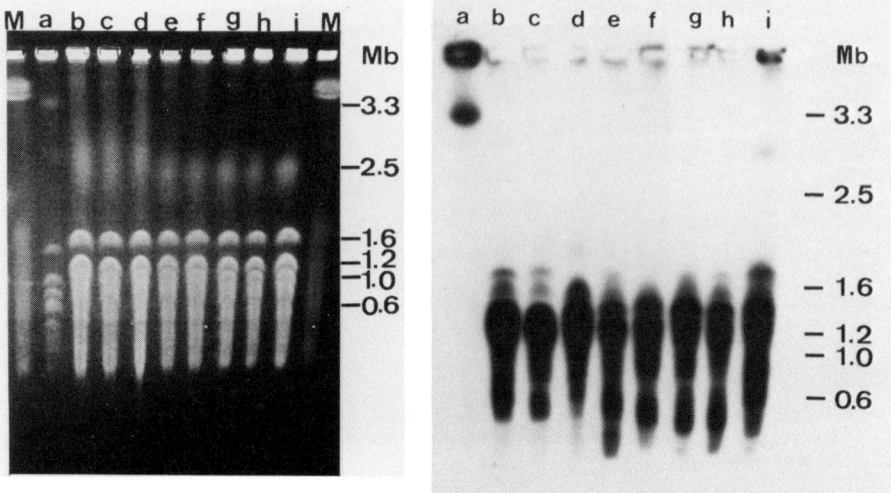

Figure 11. Eight agarose plugs each containing DNA from 768 YAC clones (8 × 96) analysed using the CHEF 'Pulsaphor' apparatus (Pharmacia). The 1% (w/v) agarose gel (SeaKem, FMC) in 0.5 × TBE was subjected to electrophoresis at 70 V with a 15 min pulse time, for 117 h at 12 °C. Lane M represents chromosome markers from *S. pombe*. Lane a represents a mouse genomic YAC (3.3 Mb). (a) Photograph of gel stained with ethidium bromide; (b) the corresponding transfer filter probed with ^{32}P-dCTP labelled pBR322.

The construction of an available megabase insert human genome YAC library will allow an increase in the assembly of large contigs using fewer ordered clones with the coincidence of cloned genome regions delimited by STS (Sequence Tagged Sites), polymorphic landmarks 1 cM: centiMorgan (~10^6 bp) apart, and the isolation of chromosome-specific YAC libraries (17–19).

Acknowledgements

We would like to thank I. M. Chumakov, M. James, and C. Stark for helpful reading of the manuscript, and A. Billault, M. Saumier and V. Perrot for their advanced technical assistance. This work was supported by the M.R.E. (The French ministry of research and space) no. 92S0061 and by A.F.M. (Association Francaise contre les Myopathies).

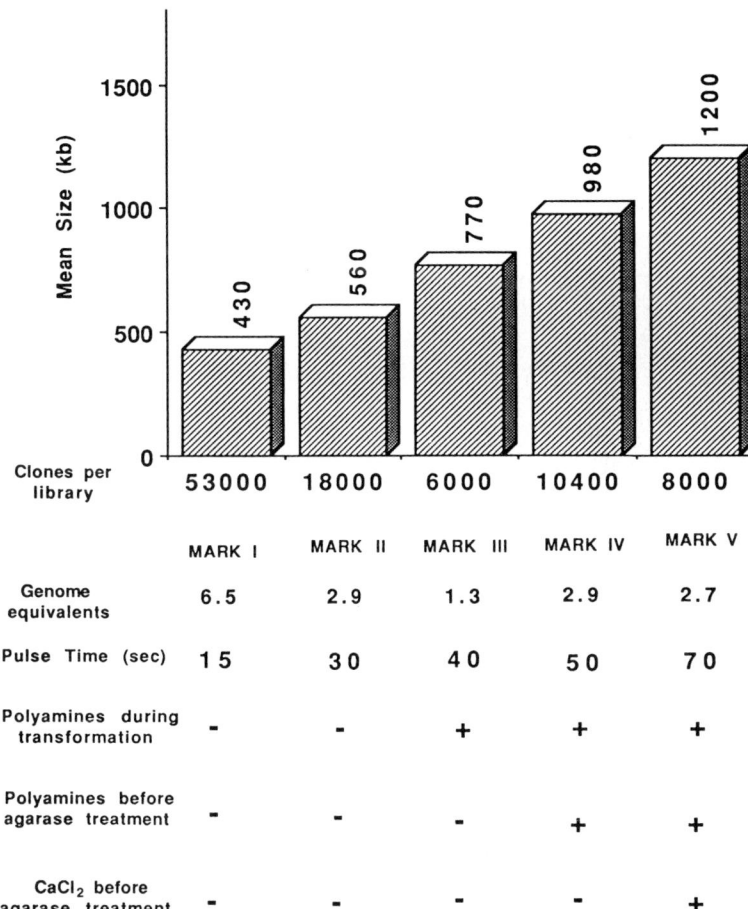

Figure 12. Representation of the CEPH YAC library evolution. The modifications (pulse time, polyamines, and CaCl$_2$) included at different crucial steps of the classical YAC cloning protocol are indicated for each successive library (−, no modification +, modification). The combination of all modifications have produced the last generation of clones (MARK V) also called the CEPH 'Mega-YAC' library.

References

1. Burke, D. T., Carle, G. F., and Olson, M. V. (1987). *Science*, **236**, 806.
2. Anand, R., Riley, J. H., Butler, R., Smith, J. C., and Markham, A. F. (1990). *Nucleic Acids Res.*, **18**, 1951.
3. Anand, R., Villasante, A., and Tyler-Smith, C. (1989). *Nucleic Acids Res.*, **17**, 3425.
4. Imai, T. and Olson, M. V. (1990). *Genomics*, **8**, 297.
5. Albertsen, H. M., Abderrahim, H., Cann, H., Dausset, J., Le Paslier, D., and Cohen, D. (1990). *Proc. Natl Acad. Sci. USA*, **87**, 4256.

6. Larin, Z., Monaco, A. P., and Lehrach, H. (1991). *Proc. Natl. Acad. Sci. USA*, **88**, 4123.
7. Schwartz, D. C. and Cantor, C. R. (1984). *Cell*, **37**, 67.
8. Chu, G., Vollrath, D., and Davis, R. W. (1986). *Science*, **234**, 1582.
9. Chu, G. (1989). *Electrophoresis*, **10**, 290.
10. Birren, B. W., Lai, E., Clark, S. M., Hood, L., Simon, M. I. (1988). *Nucleic Acids Res.*, **16**, 7563.
11. Burke, D. T. and Olson, M. V. (1990). In *Methods in enzymology* (ed. C. Gutherie and G. R. Finks), Vol. 194, pp. 251–73. Academic Press, San Diego.
12. Carle, G. F., Frank, M., and Olson, M. V. (1986). *Science*, **232**, 65.
13. Chartier, F. L., Keer, J. T., Sutcliffe, M. J., Henriques, D. A., Milham, P., and Brown, S. D. M. (1992). *Nature Genet.*, **1**, 132.
14. Cornelly, C., McCormick, M. K., Shero, J., and Hieter, P. (1991). *Genomics*, **10**, 1.
15. Burger, P. M. J. and Percival, K. J. (1987). *Anal. Biochem.*, **163**, 391.
16. Dausset, J., Ougen, P., Abderrahim, H., Billault, A., Sambucy, J.-L., Cohen, D., and Le Paslier, D. (1992). *Behring Inst. Mitt.*, **91**, 13.
17. Chumakov, I. M., Rigault, P., Guillou, S., Ougen, P., Billault, A., Guasconi, G., *et al.* (1992). *Nature Genet.*, **1**, 38.
18. Chumakov, I. M., Le Gall, I., Billault, A., Ougen, P., Soularue, P., Guillou, S., *et al.* (1992). *Nature Genet.*, **1**, 222.
19. Bellanne-Chantelot, C., Lacroix, B., Ougen, P., Billault, A., Beaufils, S., Bertrand, S., *et al.* (1992). *Cell*, **70**, 1059.

6

PFGE analysis of yeast artificial chromosomes

J. RAGOUSSIS

1. Introduction

There are two basic pulsed field gel electrophoresis (PFGE)-related techniques for mapping yeast artificial chromosomes (YACs). One is to digest the DNA with restriction endonucleases, detect specific fragments with defined probes, and construct a restriction map. The other technique is to generate YAC variants—(so called fragmentation products)—by homologous recombination with vectors containing a repeated (e.g. Alu), or specific, sequence. Using the recombination potential of yeast allows the construction of YAC derivatives for mapping of sequences within a YAC.

2. Restriction analysis of YACs

2.1 Introduction

Many YACs are isolated using a single-copy specific probe or sequence tagged site (STS) from a region lacking additional markers. This limits the extent of subsequent restriction mapping of the clone. The linear nature of a YAC is an advantage in such cases, since it allows the detection of partial-digestion restriction fragments by using probes specific for the vector sequences present at the two ends. This circumvents the need for specific internal probes. However, when available, specific internal probes can be mapped within the YAC using single and double restriction enzyme complete digests.

2.2 Selection of enzymes

Rare-cutting, methylation-sensitive endonucleases are widely used for long-range mapping (see Chapter 2). Since YAC DNA is not methylated, these enzymes are expected to cut the cloned DNA more frequently than genomic DNA. Rare-cutting restriction enzymes can be divided into those which recognize CpG islands with high frequency, for example: *Not*I, *Bss*HII, and *Eag*I and those that do so with lower frequency such as: *Mlu*I, *Nru*I, and *Pvu*I (1, 2).

A combination of the two classes of rare-cutting restriction enzymes is ideal for generating physical maps of YAC clones. The resulting maps also provide information about the position of CpG islands and thus potentially associated genes (3).

2.3 Preparation of agarose plugs

The plugs can be prepared by following one of two lysis methods described in *Protocol 1*. The protocols are modified versions of those described in the literature (4, 5). The lithium lysis method is simpler, less expensive, and very effective.

Protocol 1. Preparation of high molecular weight yeast DNA in agarose plugs

Equipment and reagents

- SDcaa: per litre, 6.7 g yeast nitrogen base (Difco) + 14 g casamino acids (Difco) + 50 ml 40% (w/v) glucose (filter, sterilize, and add after autoclaving)
- Selective medium (SD): per litre, 1.7 g yeast nitrogen base without amino acids and without (NH$_4$)$_2$SO$_4$ (Difco) + 5.0 g (NH$_4$)$_2$SO$_4$/amino acids without the one(s) to be used for selection (see below) + 2% (w/v) glucose, pH 5.8. Final concentrations of amino acids:
  adenine (20 μg/ml)
  arginine (20 μg/ml)
  isoleucine (20 μg/ml)
  histidine (20 μg/ml)
  leucine (60 μg/ml)
  lysine (20 μg/ml)
  methionine (20 μg/ml)
  phenylalanine (50 μg/ml)
  tryptophan (20 μg/ml, light sensitive, filter and sterilize)
  valine (150 μg/ml)
  tyrosine (30 μg/ml, needs NaOH to go into solution)
  uracil (20 μg/ml)
- SD/agar plates: add 17–20 g/litre bactoagar (Difco) to SD medium before autoclaving
- SE: 1 M sorbitol + 20 mM EDTA
- Low melting temperature (LMT) agarose
- 2-mercaptoethanol
- SET: 1 M sorbitol + 20 mM EDTA + 10 mM Tris–HCl pH 7.5
- Lyticase (Sigma, 50 units/μl in SE) or Zymolyase 100T (ICN Biomedicals)
- Perspex moulds with slots of approximately 100 μl volume (available from BioRad or Pharmacia Biotech)
- Pasteur pipette bulb
- YLS: 1% (w/v) lithium dodecylsulphate + 100 mM EDTA pH 8.0 + 10 mM Tris–HCl pH 8.0 (YLS is very toxic, take care and wear gloves)
- 400 mM EDTA, pH 7.5 + 1% (w/v) N-laurylsarcosine + 2 mg/ml Proteinase K
- TE: 10 mM Tris–HCl pH 8.0 + 1 mM EDTA pH 8.0
- 0.5 M EDTA pH 8.0
- Shaking incubator (New Brunswick Scientific)
- Medium range centrifuge (Beckman J6-B)
- Sterile pipette tips, 50 ml plastic tubes

Method

1. Inoculate a single colony into 3–20 ml Ura$^-$, Trp$^-$ media (SD or SDcaa). Approximately 1–2 plugs or 1–2 μg of DNA will be obtained per ml of saturated culture. Grow for 24–48 h at 30°C with shaking.

2. Pellet the cells by centrifugation at 1000 *g* for 5 min and resuspend the pellet in 20 ml of 50 mM EDTA, pH 8.0.

3. Dissolve LMT agarose to a concentration of 2% (w/v) in SE. Bring to 45°C and store at this temperature. Add 2-mercaptoethanol to 14 mM.

---

4. Pellet the cells by centrifugation at 1000 *g* for 5 min. Resuspend the cells in 400 μl SE + 14 mM 2-mercaptoethanol + 100 mg/ml Zymolyase 100T or 20 units/ml lyticase.

5. Prepare plug moulds by briefly rinsing in ethanol, drying, and taping up the bottom. Add 400 μl of the molten agarose to the resuspended cells, mix well, and dispense into the slots of the plug mould. Allow to set on ice for 20 min.

6. Expel the plugs by placing a pasteur pipette bulb over the bottom of the plug in the mould and squeezing firmly whilst holding over a 30 or 50 ml plastic tube containing 5 ml SET + 14 mM 2-mercaptoethanol + 100 mg/ml Zymolyase 100T or 20 units/ml Lyticase. Incubate at 37°C for 2 h (shake gently or invert the tube a couple of times carefully).

7. After spheroplasting the yeast cells in agarose plugs, there are two methods to lyse the cells.

   (a) Lithium lysis method. Decant the spheroplasting solution and replace with 5 ml of filtered, sterilized YLS. Incubate at 37°C for 30 min with gentle shaking. Pour off the YLS solution and add another 5 ml YLS. Incubate overnight at 37°C. The changing of solutions is made easy with a special sieve-cup for 50 ml tubes, available from BioRad.

   (b) Proteinase K lysis method. Transfer the plugs to 400 mM EDTA, pH 7.5 + 1% (w/v) *N*-laurylsarcosine + 2 mg/ml Proteinase K. Incubate at 50°C for 24–48 h.

8. Store the plugs in yeast lysis solution at room temperature or rinse and store in 500 mM EDTA at 4°C.

2.4 Sizing of YACS using PFGE

In order to determine the size of YACs separate them by PFGE as outlined in *Protocol 2* (also see Chapter 5, Section 9.2). They are often visible as extra bands compared to the natural yeast host chromosomes. If not visible on the ethidium-bromide stained gel, detect them by Southern blot hybridization using pYAC4 vector or species-specific repeats as hybridization probes. This is a necessary step for the identification of additional YACs in the same clone.

Protocol 2. Separation of YAC DNA by PFGE

Equipment and reagents

- 0.5 × TBE electrophoresis buffer: 10 × TBE stock solution per litre: 108 g Tris base + 58 g boric acid + 40 ml 0.5 M EDTA pH 8.0
- TE (see *Protocol 1*)
- T0.1E: 10 mM Tris–HCl pH 8.0 + 0.1 mM EDTA pH 8.0
- 0.25 M HCl
- 0.4 M NaOH
- PFGE apparatus (BioRad, Pharmacia Biotech, or Biometra)
- Vacuum blotting apparatus, optional (BioRad, Pharmacia, Hybaid and others)

Protocol 2. *Continued*

- Blotting membrane (Hybond, Amersham or others)
- 1% (w/v) agarose in 0.5 × TBE
- Ethidium bromide (2 μg/ml)

- 5 × SSC: 20 × SSC stock solution per litre: 175.3 g NaCl + 88.2 g sodium citrate, pH 7.0 with NaOH

A. *Preparation of plugs and electrophoretic separation*

1. Wash the plugs thoroughly 3 times for 30 min each in TE at 50°C and 3 times for 30 min each in T0.1E at room temperature.

2. Prepare a 1% (w/v) agarose gel in 0.5 × TBE.

3. Load plugs and separate under the following conditions (for BioRad CHEF-DRII apparatus):

   (a) for YACs obtained from large insert libraries (e.g. CEPH or ICRF): ramped pulse time 40–90 sec; 6 V/cm; 24 h;

   (b) for YACs larger than 900 kb: ramped pulse time 60–120 sec; 6 V/cm; 24–26 h;

   (c) for YACs smaller than 450 kb: ramped pulse time 10–40 sec; 6 V/cm; 24 h.

4. Stain the gel with ethidium bromide (2 μg/ml) and photograph with a ruler at the side.

B. *Preparation of a Southern blot*

Use blotting conditions according to the recommendations of the membrane manufacturer. Positively charged membranes are suitable for multiple hybridizations. Membranes suitable for alkaline blotting are easy to use. For Amersham Hybond N+ and related charged membranes proceed as follows.

1. Wash the gel in 0.25 M HCl for 20 min to depurinate the DNA.

2. Prepare a standard capillary Southern blot with 0.4 M NaOH as transfer buffer for a minimum of 4–5 h. Alternatively, use a vacuum blotting apparatus according to the manufacturer's instructions. In this way the transfer can be accomplished within 2 h.

3. Neutralize the membrane by washing for 2 min in 5 × SSC.

4. Detect the YACs as bands either by hybridization with a radioactively labelled specific DNA probe or total human DNA.

2.5 Partial digestion of YAC DNA

2.5.1 Basic preparation of partial digests

The method for partial restriction enzyme digestion of DNA prepared in agarose plugs is outlined in *Protocol 3*.

Protocol 3. Restriction enzyme digestion of DNA prepared in agarose plugs

Equipment and reagents

- LMT agarose YAC plugs prepared as in *Protocol 1*
- TE (see *Protocol 1*)
- Restriction enzyme (RE) buffer (see manufacturers' recommendations)
- Restriction enzymes (New England Biolabs and various other suppliers)
- Phenylmethylsulfonylfluoride (PMSF, 0.04 mg/ml); toxic, wear gloves
- 0.5 M EDTA pH 8.0
- Water bath
- Sterile pipette tips
- 1.5 ml microcentrifuge tubes

Method

1. Wash the LMT agarose plugs containing 3–5 µg of DNA with TE (up to five plugs in 50 ml of TE) as follows: 3 times for 1 h each at room temperature, once overnight at 4°C, and again for 1 h at room temperature using fresh TE. If the Proteinase K lysis method was used (*Protocol 1*) wash the plugs in TE twice at 50°C, TE with PMSF (0.04 mg/ml) twice at 50°C (to inactivate the Proteinase K), and TE twice at room temperature.

2. Equilibrate each plug in 1 ml of RE buffer for 1 h on ice.

3. Place into a fresh 1.5 ml microcentrifuge tube containing 300 µl of RE buffer plus restriction enzyme for one plug. At least three different amounts of enzyme are recommended (see below) plus one undigested and one complete digest control.

4. Equilibrate for 1 h on ice and then incubate for 15 min at 37°C (or 50°C for *Bss*HII digests), except for *Not*I digests where a 1 h incubation is optimal. Stop the reactions by adding 50 ml of 0.5 M EDTA pH 8.0 and place the tubes on ice (see *Figure 1*). The units of each enzyme to be used in the individual digests is given below. Please note that all our digests have been performed by using enzymes from New England Biolabs. Slight adjustments may be necessary when using enzymes supplied by other companies.

   (a) For *Not*I partial digests: 1, 2, and 3 units of enzyme (note: 1 h incubation at 37°C) and for the complete digest use 25 units for 2 h at 37°C.

   (b) For *Bss*HII partial digests: 0.1, 0.2, 0.4, 0.6, or 1 units; for 15 min at 50°C and for the complete digest use 12.5 units for 1 h at 50°C.

   (c) For *Mlu*I partial digests: 1, 5, and 10 units for 15 min at 37°C and for the complete digest use 50 units for 1 h at 37°C.

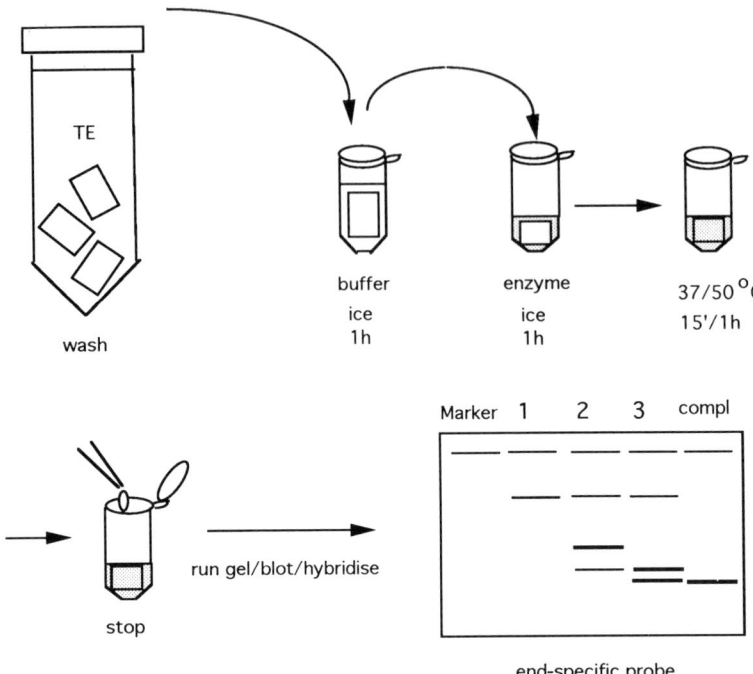

Figure 1. Schematic diagram of a partial digestion experiment. The agarose plugs containing YAC DNA are equilibrated in RE buffer, then with buffer containing enzyme on ice, and then digested at the optimal temperature for a short time. Hybridization of a Southern blot of PFGE-separated digestion products with vector-end specific probes reveals a ladder of bands. 1, 2, 3 represent increasing numbers of enzyme units and complete digestion (compl).

After digestion separate the DNA by PFGE using conditions which correlate to the size of the YAC and the PFGE apparatus. For example, in short gels (less than 15 cm available running distance) fragments smaller than 5 kb may run off the end of the gel. Special conditions may also be used to open the separation window over an area containing a cluster of digestion sites. All this is illustrated in the examples given in Section 2.5.3.

2.5.2 End-specific probes

To detect the ends of the YAC use probes derived from pBR322 DNA. Digestions with *Pvu*II and *Bam*HI result in a 2.67 kb fragment specific for the 'left' end (containing Trp and centromere) and a 1.69 kb fragment specific for the 'right' (Ura) end. Alternatively a *Pvu*I and *Eco*RI double digest releases a 0.68 kb fragment, which can be used as left-end-specific probe while a 1.4 kb *Pvu*II and *Sal*I double digest product is suitable as a right-end-specific probe.

Vector end specific probes can be derived by PCR using pYAC4 DNA (6). A 330 bp left-arm-specific probe is made by amplification with the primer pairs YACLR1 (5'-GTGTGGTCGCCATGATCGCG-3') and YACLP (5'-ATGCGGTAGTTTATCACAGTTAA-3'). The 265 bp long right-arm probe is amplified with the primer pair YACRP (5'-GATCATCGTCGCGC-TCCAAGCGAAAGC-3') and YACRR3 (5'-CTCGCCACTTCGGGCTCA-3'). PCR is carried out for 30 cycles using standard reaction buffers containing 1.5 mM MgCl$_2$ and an annealing temperature of 55°C (primers and conditions are as suggested by the authors, ref. 6).

2.5.3 Examples of mapping YAC clones by partial digestion

i. Mapping of a 450 kb YAC using BssHII and MluI partial digests

The partially digested DNA was separated on a 1% (w/v) agarose gel under the following conditions: 3–30 sec ramped pulse time at 200 V, 20 h (*Figure 2*) or 3–30 sec ramped pulse time at 210 V, 18 h. The following fragment sizes (given in kb) were determined (*Figure 2*):

Left end		Right end	
*Bss*HII	*Mlu*I	*Bss*HII	*Mlu*I
23	35	150	20
35	80	160	170
150	150	170	320
280	300	220	

To allow for better resolution over the 50–250 kb range the following conditions have been applied: pulse time 0.3–30 sec at 210 V, 19 h. Thus, we were able to detect three *Bss*HII sites within 50 kb from each other, suggesting the position of a CpG island (*Figure 3*).

The restriction sites can be drawn on a line with the sites detected by the left probe above and the sites detected by the right probe below the line. In general, there is agreement between the positions estimated by calculating the distances from left and right. In the case of a discrepancy the result obtained with the probe closer to the particular end is the most reliable (see *Figure 4*).

ii. Mapping of a 1050 kb long YAC using BssHII partial digests

The 1050 kb YAC was partially digested with 0.2, 0.6, and 0.1 units of enzyme and the DNA separated on a BioRad CHEF-DRII apparatus

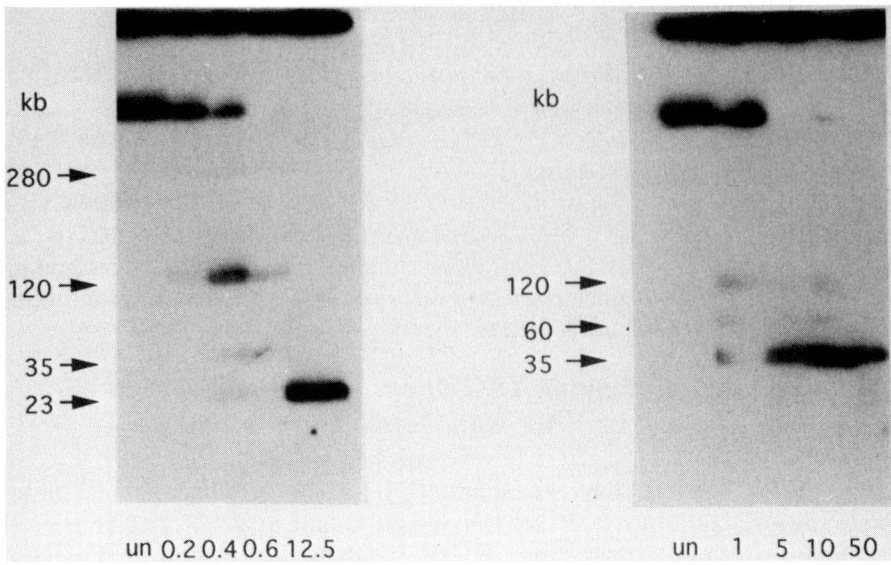

Figure 2. Partial digestion of a 450 kb YAC with *Bss*HII and *Mlu*I. The number of units used are shown at the bottom of each gel. Running conditions: 3–30 sec ramped pulse time at 200 V, 20 h. The blot was hybridized with the left YAC arm-specific probe.

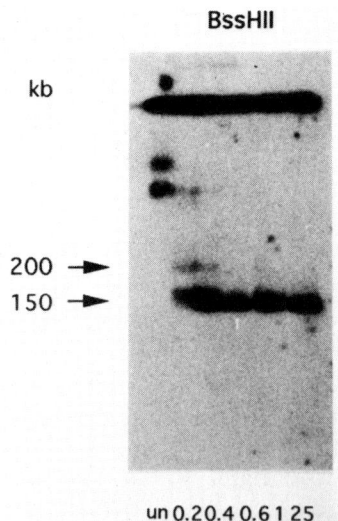

Figure 3. The same YAC as in *Figure 2* was partially digested with *Bss*HII and the fragments separated on a gel under the following conditions: pulse time 0.3–30 sec at 210 V, 19 h. The number of units are shown beneath the gel. The blot was probed with the right YAC arm-specific probe. Under these conditions three bands are separated around 150 kb, representing a cluster of *Bss*HII sites.

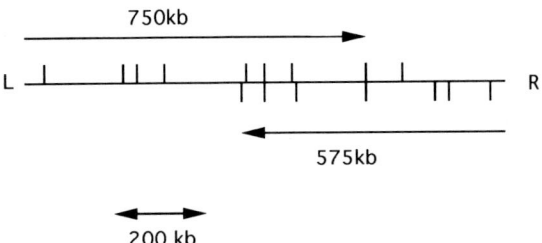

Figure 4. Construction of the 450 kb YAC map for *Bss*HII and *Mlu*I. The sites detected by the left probe are placed above the line, the ones detected by the right probe placed underneath. The positions of most sites are nearly identical whether detected by the one or the other probe, particularly for *Mlu*I. The *Bss*HII site not seen with the left probe is shown, indicating the importance of using both end-specific probes.

under the following conditions: 200 V, 40–90 sec ramped pulse time, for 20 h. Fragments detected (*Figure 5*):

Left-arm probe (kb)	Right-arm probe (kb)
49	36
218	121
243	146
315	291
490	450
533	525
582	575
750	
820	

The map resulting from this experiment is illustrated in *Figure 6*. It is obvious that the accuracy in such a range of separation is limited. For example, there may be multiple restriction sites in the 500–600 kb region. If necessary, it is worth saving one half of the partially digested plugs in order to load a second gel and apply conditions allowing a finer separation over a particular size range.

2.6 Complete digests and double digests

2.6.1 Complete digests

The method outlined in *Protocol 3* for partial digests can be used also for complete digests with any restriction endonuclease. *Protocol 4* outlines the method used for digesting DNA prepared in agarose plugs with two different restriction enzymes.

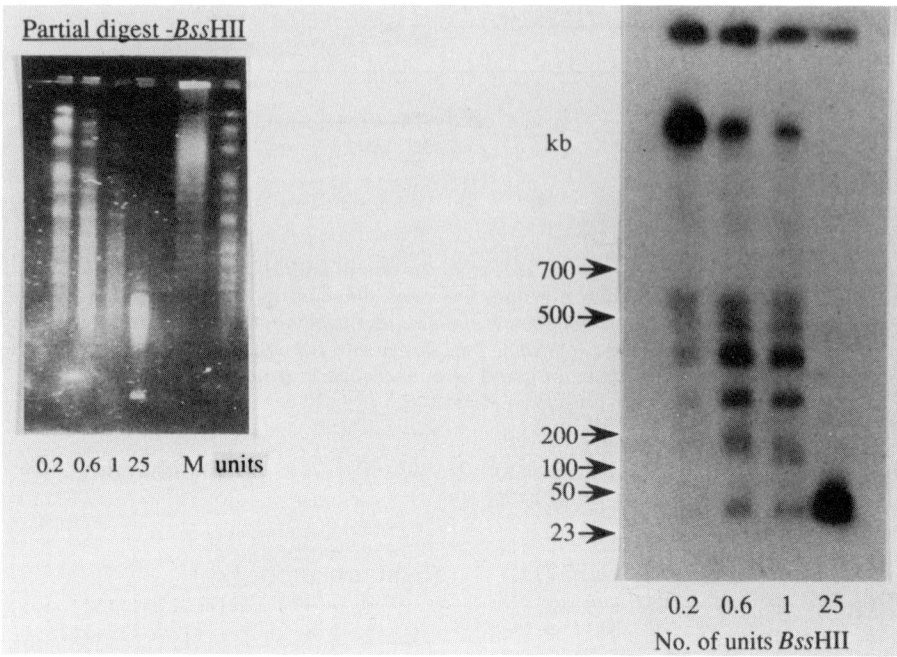

Figure 5. Partial digestion of a 1050 kb YAC with *Bss*HII. The ethidium bromide-stained gel is on the left. The progression from partial (0.2 units) to complete digestion is visible with increasing number of units used. Hybridization with the right-arm probe reveals a ladder of bands up to approximately 600 kb.

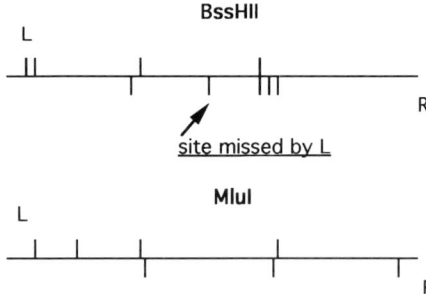

Figure 6. Construction of the 1050 kb map. The positions of the *Bss*HII sites detected is shown. The definite positions of sites up to 750 kb from the one end was determined with the left YAC-end probe and 575 kb from the other with the right YAC-end probe.

2.6.2 Double digests

Protocol 4. Double restriction enzyme digestion of DNA prepared in agarose plugs

Equipment and reagents

See *Protocol 3.*

Method

1. Digest one plug with the first enzyme and cut it in half.

2. Store one half of the plug in TE at 4°C.

3. Use the other half of the plug for the second digest. If the same RE buffer is suitable for both enzymes simply incubate the half plug in 250 µl fresh buffer containing the enzyme. If a different buffer is optimal for the second enzyme pre-incubate the half plug in 500 ml RE buffer for 1 h at 4°C and proceed as described above.

4. Incubate at the appropriate temperature for 2–16 h. Minimum units for a selection of enzymes:

*Bss*HII:	20U
*Cla*I:	20U
*Eag*I:	10U
*Mlu*I:	25U
*Not*I:	10U
*Pvu*I:	25U
*Sal*I:	20U
*Xho*I	25U

2.7 PFGE markers

A selection of markers are recommended such as λ-concatemers and yeast chromosomes available from Pharmacia, BioRad, New England Biolabs, and other companies. Also recommended are the mid-range PFGE markers (24–600 kb) available from New England Biolabs. See Chapter 2, Section 4.2 and *Table 2* for a more detailed discussion.

In order to generate your own low-range combination you can mix 100 ng of λ *Hind*III-digested DNA and 100 ng λ *Sal*I-digested DNA with 100 µl melted 1% (w/v) agarose in 0.5 × TBE and poured into a mould. The resulting plug can be easily loaded on PFGE gels.

3. YAC fragmentation

3.1 Introduction

A complementary method of mapping sequences within a YAC is the generation of deletion derivatives. This technique, like many others, takes advantage of the wide range of possibilities to manipulate DNA in yeast by homologous recombination. The YAC can be deleted from the right (acentric) or left end (centric) by recombination between repetitive sequences present in the YAC and those present in special vectors. These plasmid constructs contain an Alu or LINE repetitive sequence, a selectable marker, and a telomere. A centromere is added for left-end specific deletions. A range of constructs have been published (7–10).

YAC fragmentation has been used to map highly homologous gene family members within a YAC (11) which avoids cross-hybridization problems between related genes. Furthermore, it can be used to map STS markers along the YAC by simple PCR analysis of each STS using DNA isolated from individual fragmentation clones.

One set of constructs is available from Dr Roger Reeves, Department of Physiology, Johns Hopkins University School of Medicine, Baltimore, MD 21205, USA. It comprises the pBP series of vectors which contain HIS3 as the selectable marker and an Alu sequence in different orientations (pBP108 and pBP109) or a LINE sequence (pBP110 and pBP111). In addition, the construct pBP81 contains a centromere and two head-to-tail Alu sequences allowing fragmentations from the YAC left end.

Another Alu repeat sequence fragmentation vector, pBCL8.1, has been reported by Dr Christopher Denny, Department of Pediatrics, A2-312 MDCC, UCLA Medical Center, LA CA 90024, USA (10). It is available from the American Type Culture Collection (ATCC, Rockville, Maryland, USA). It has the advantage of using an auxotrophic marker (*LYS2*) compatible with the AB1380 host, but it can only be used for acentric fragmentation.

3.2 Fragmentation methods

3.2.1 The pBP series of fragmentation vectors

To allow for histidine selection the YAC must be transferred from the commonly used AB1380 host to a His3D200 host such as YPH252 (haploid host) or YPH274 (diploid host). This can be accomplished by mating with a suitable haploid host, sporulation, and tetrad dissection. Another possibility is to transform the YAC into the appropriate strain; this is outlined in *Protocol 5*. This enables the use of the fragmentation vectors without any experience in tetrad dissection. The spheroplast transformation protocol is a modification of Burgers and Percival (12).

Protocol 5. Transformation of YAC DNA into *S. cerevisiae* strains

Equipment and reagents

- Sterile distilled water
- TEN + polyamines: 10 mM Tris–HCl pH 7.5 + 1 mM EDTA + 30 mM NaCl + 0.75 mM spermidine trihydrochloride (Sigma) + 0.3 mM spermine tetrahydrochloride (Sigma)
- YPD: per litre, 10 g yeast extract (Difco) + 20 g bactopeptone (Difco) + 2% (w/v) glucose, pH 5.8. Add bactoagar (Difco) at 2% (w/v) for YPD agar plates.
- Regeneration agar for top/plates: 1 M sorbitol + 2% (w/v) glucose + 0.67% (w/v) yeast nitrogen base + amino acids (see *Protocol 1*, −Ura) + 2% (w/v) agar
- SCE: 1 M sorbitol + 0.1 M sodium citrate pH 5.8 + 10 mM EDTA pH 7.5 + 30 mM 2-mercaptoethanol (add fresh)
- STC: 1 M sorbitol/10 mM $CaCl_2$/10 mM Tris–HCl pH 7.5
- PEG: 20% (w/v) polyethylene glycol (PEG 6000, Sigma or PEG 4000, Serva) + 10 mM $CaCl_2$ + 10 mM Tris–HCl pH 7.5

- SOS: 1 M sorbitol + 25% (w/v) YPD + 6.5 mM $CaCl_2$ + 1 mg/ml uracil (or amino acid according to the selectable marker present in the YAC)
- Sodium lauryl sulphate (SDS), stock solution 10% (w/v)
- Beta-agarase I (New England Biolabs)
- Lyticase (Sigma)
- Medium range centrifuge (Beckman J6-B)
- Shaking incubator (New Brunswick Scientific)
- Incubator (LEEC)
- Spectrophotometer
- Haemocytometer
- Standard light microscope with phase contrast
- Water bath
- Sterile pipette tips, 15 ml and 50 ml plastic tubes, sterile Petri dishes, and 1.5 ml microcentrifuge tubes

A. *Preparation of DNA*

1. Prepare the agarose plugs as described in *Protocol 3* and wash them thoroughly with TE. The plugs can be made more concentrated as described in Chapter 7, *Protocol 3*.

2. Wash four times for 30 min in TEN + polyamines.

3. Melt a plug in a 1.5 ml microcentrifuge tube at 68°C, bring to 40°C for 10 min, and add 20 units of beta-agarase I. Incubate for 1–2 h at 40°C until the agarose is completely digested.

B. *Spheroplast transformation*

1. Streak a fresh YPD agar plate with the strain to be transformed from a frozen glycerol stock (for example YPH252). Grow at 30°C for 2–3 days. Inoculate a single colony into 10 ml of YPD and shake overnight at 30°C.

2. On the next evening, inoculate 200 ml of YPD in a 1000 ml flask with 1 ml of the 10 ml overnight culture (step 1). Shake at 30°C overnight.

3. Split the culture into 50 ml plastic tubes (when using a spectrophotometer the $OD_{600 nm}$ of a one-tenth dilution is between 0.1 and 0.15). Test some of the culture under the microscope for bacterial contamination. Pellet the cells by centrifugation at 1000 g for 5–10 min at 20°C.

4. Decant the media and resuspend the pellets in 20 ml of sterile distilled water for each tube. Pellet the cells by centrifugation at 1000 g for 5–10 min at 20°C.

Protocol 5. *Continued*

5. Decant the water and resuspend the pellets in 20 ml of 1.0 M sorbitol. Pellet the cells by centrifugation at 1000 g for 5–10 min at 20°C.

6. Decant the sorbitol and resuspend the pellets in 20 ml SCE. Add 46 μl of 2-mercaptoethanol and take out 300 μl from one tube for a pre-lyticase control. Add 500–1000 units of lyticase, mix gently, and incubate at 30°C.

7. At 5, 10, 15, and 20 min, test the extent of spheroplasting of one tube by two independent methods.

   (a) On the spectrophotometer measure the $OD_{600\ nm}$ of a one-tenth dilution in 5% (w/v) SDS. When the value is 1/10 of the pre-lyticase value, spheroplasting is >90% complete.

   (b) Mix 10 μl of cells with 10 μl of 2% (w/v) SDS and check under the microscope using phase contrast. When cells are dark they are spheroplasted. Take the spheroplasting to 90%. This should take 15–20 min. At this point move quickly, as the cells should not be over-spheroplasted. Pellet the cells by centrifugation at 150–300 g for 5 min at 20°C.

8. Decant the SCE and gently resuspend the pellets in 20 ml of 1.0 M sorbitol. Pellet the cells by centrifugation at 150–300 g for 5 min at 20°C.

9. Decant the sorbitol, and resuspend the pellets in 20 ml STC. Take a cell count of one tube by making a 1/10 to 1/50 dilution in STC and count the number of cells on a haemocytometer. Pellet the cells by centrifugation at 150–300 g for 5 min at 20°C and resuspend the cells in the volume of STC calculated for the desired final concentration (4.5–6×10^8 cells/ml).

10. Prepare 15 ml clear polystyrene tubes with up to 50 μl of carefully pipetted YAC DNA (use cut pipette tips). Add 150 μl of spheroplasted cells in STC. Let the DNA and cells sit for 10 min at 20°C.

11. Add 1.5 ml of PEG and mix gently by inverting the tubes. Let sit for 10 min at 20°C. Pellet the cells by centrifugation at 150–300 g for 8 min at 20°C.

12. Carefully pipette off the PEG and do not disturb the pellets. Gently resuspend the pellets in 225 μl SOS. Place at 30°C for 30 min.

13. Add 8 ml of selective regeneration top agar at 48°C. Mix gently and pour quickly on to the surface of a pre-warmed selective (Ura⁻) regeneration plate and allow to set. Incubate the plates upside down at 30°C for 3–4 days.

This protocol should provide enough transformants and even one is enough! Now the YAC is in the appropriate yeast strain and can be transformed with the fragmentation vectors as outlined in *Protocol 6*. The protocol for lithium acetate transformation of yeast is a modification from Gietz *et al.* (13).

Protocol 6. YAC fragmentation

Equipment and reagents

- Sterile distilled water
- SD medium without uracil (see *Protocol 1*)
- SD medium agar plates without histidine (see *Protocol 1*)
- TE (*Protocol 1*) + 100 mM lithium acetate pH 7.5
- Sonicated and denatured salmon sperm DNA
- TE + 40% (w/v) PEG 1500 (BDH Laboratory Supplies) + 100 mM lithium acetate pH 7.5

- Medium range centrifuge (Beckman J6-B)
- Microcentrifuge
- Shaking incubator (New Brunswick Scientific)
- Incubator (LEEC)
- Water bath
- Sterile pipette tips, 15 ml and 50 ml plastic tubes, 1.5 ml microcentrifuge tubes, and sterile Petri dishes

A. *Preparation of the vector*

Linearize the fragmentation vector with the appropriate restriction enzyme (*Sal*I for Alu fragmentation) using restriction enzyme buffers and temperatures recommended by the manufacturer.

B. *Lithium acetate transformation*

1. Grow an overnight culture (10 ml) at 30°C with shaking and on the next morning dilute the culture by 1/10th into fresh medium (SD/Ura$^-$, 50 ml) and grow for 4 h.

2. Pellet the cells by centrifugation (1000 *g*, 5 min) and resuspend in sterile distilled water. Pellet the cells again.

3. Resuspend the cells in 1 ml water and transfer to a 1.5 ml microcentrifuge tube.

4. Pellet the cells by spinning briefly in a microcentrifuge.

5. Wash the cells in 1 ml TE + 100 mM lithium acetate (pH 7.5).

6. Resuspend the cells in 1 ml TE + 100 mM lithium acetate (pH 7.5).

7. Mix together 50 μl yeast + 1 μg DNA + 50 μg salmon sperm DNA.

8. Add 300 μl of 40% (w/v) PEG 1500 in TE + 100 mM lithium acetate (pH 7.5) and mix.

9. Incubate at 30°C for 30 min while shaking.

10. Heat-shock at 42°C for 15 min and spin for 5 min in a microcentrifuge.

11. Resuspend the cell pellet in 1 ml SDcaa containing 20 μg/ml histidine and incubate at 30°C for 30 min to 1 h.

Protocol 6. *Continued*

12. Pellet the cells in a microcentrifuge and resuspend in 1 ml TE. Spread 100 μl on the surface of each plate containing SD/His⁻.

C. *Selection of transformants*

1. For pBP108–111: to determine which clones are now Trp⁺, Ura⁻, His⁺, streak out transformants on two separate His⁻, Trp⁻ SD plates with and without uracil. Colonies unable to grow on Ura⁻ plates are likely to result from a fragmentation event.

2. For pBP81 streak out on His⁻, Ura⁻ SD plates with and without tryptophan. Colonies growing only on plates without Trp are likely to contain a fragmentation product.

Note: The procedure outlined in *Protocol 5* can be used instead of the lithium acetate method. Plate out transformed cells on regeneration plates without histidine and grow for 3–4 days at 30°C (His⁺ phenotype will grow).

We found that transformations with pBP108/109 are the most efficient, while pBP81 gave very few fragmentation products per μg vector used, probably due to the nature of the construct, which contains an autonomous replicating sequence (ARS). The fragmented YAC clones can be analysed further by PFGE following *Protocol 1* and *Protocol 2*.

3.2.2 The pBCL fragmentation vector

The pBCL vector can be used for acentric fragmentation in the AB1380 host. Proceed as described in *Protocol 6*; the only difference is that the selection media is Lys⁻. After 5 days plate out the colonies in duplicate on Lys⁻, Trp⁻ plates with and without uracil. Colonies unable to grow on Ura⁻ plates are likely to result from a fragmentation event.

3.3 Examples of YAC fragmentation

3.3.1 Fragmentation of a 450 kb YAC

The YAC was transferred to yeast strain YPH274 and transformed with pBP109 as described in *Protocol 5* and *Protocol 6*. The transformation efficiency was of the order of 50–100 colonies/μg pB109. Of these, 90% had the correct phenotype (His⁺, Trp⁺, Ura⁻). Plugs were prepared from all the transformants and separated by PFGE according to *Protocol 1* and *Protocol 2*. *Figure 6* shows a typical size range of YAC fragmentation products obtained. The sizes varied from 140 kb–400 kb. Particular positions in the YAC were targeted more frequently than others, as shown in *Figure 7*. Similar results have been presented by the original investigators (7).

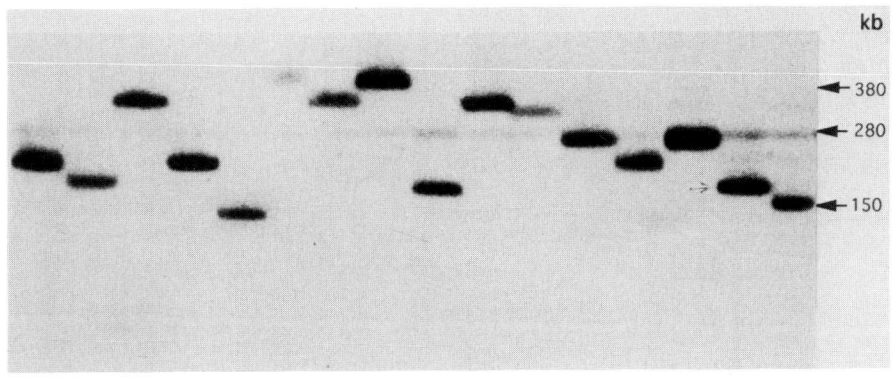

(a)

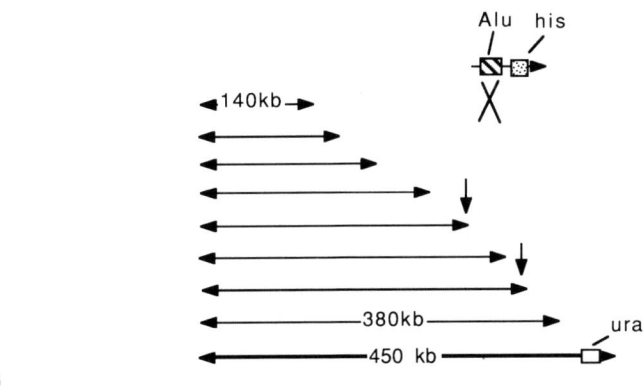

(b)

Figure 7. (a) PFGE-separated YAC fragmentation products, detected with labelled human DNA. The products were obtained with the pBP109 vector and span a size range from 140 to 380 kb. (b) The position of homologous recombination sites is illustrated in the diagram. The two vertical arrows mark the most frequent recombination events.

3.3.2 Organization of FOLR genes

The locus containing the folate receptor genes (FOLR) on chromosome 11q was cloned in a single 500 kb long YAC. This appeared to be a complex locus containing highly homologous genes. In order to study the organization of the FOLR genes, YAC fragmentation was used in conjunction with restriction enzyme mapping techniques. Clone 2.1 was fragmented as described above (Section 3.3.1), and the individual products separated by PFGE. When a Southern blot of this gel was hybridized with a general FOLR probe it was clear that only fragmentation products longer than 280 kb hybridized to the probe. This immediately positioned all cross-hybridizing genes within 120 kb from the right end of the YAC (*Figure 8*). Further analysis of the fragmentation products revealed the presence of the FOLR2L gene as the first from the left followed by FOLR1P to the right. The adult and fetal gene-specific

135

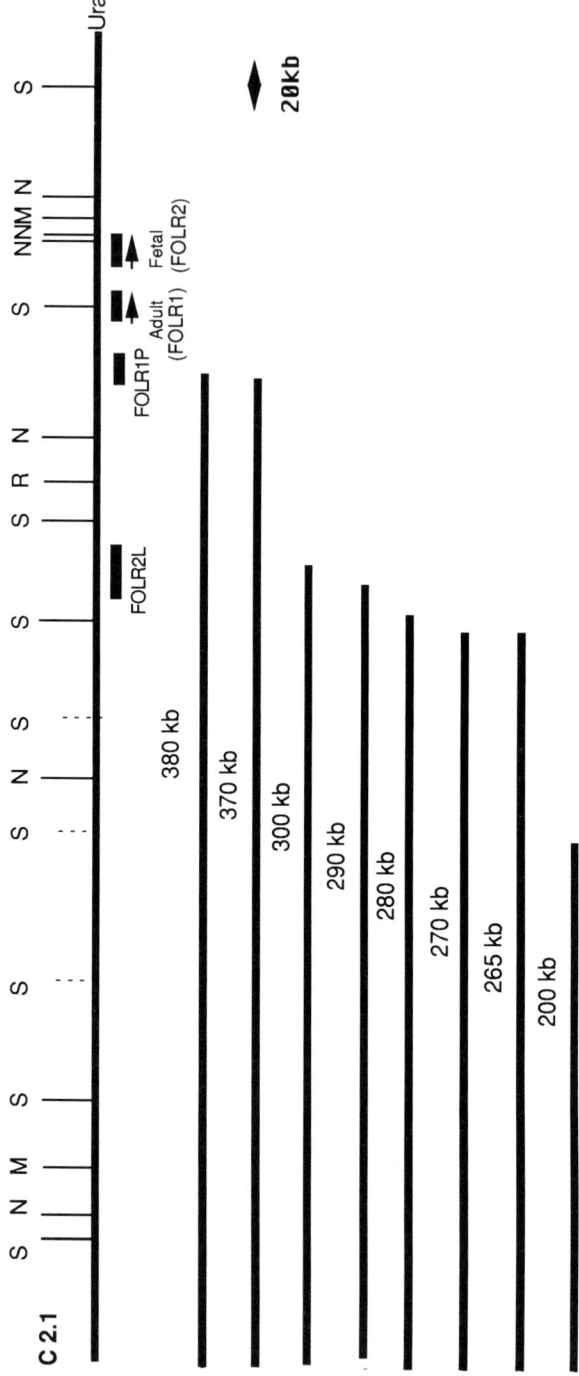

Figure 8. Application of YAC fragmentation in determining the organization of the human folate-binding protein genes. Restriction map of YAC 2.1 of 500 kb. The restriction enzyme sites shown are for *NruI* (R), *NotI* (N), *MluI* (M), and *SalI* (S). Dashed lines on the *SalI* sites indicate that the exact position of these sites has not been determined. Solid horizontal bars indicate the positions of the FOLR genes. Arrows on genes show the transcriptional orientation. The fragmented YAC clones together with their sizes, are shown below YAC 2.1. (Reproduced from ref. 11 courtesy of Academic Press.)

probes which were generated were present only in the original YAC. The physical separation of the homologous sequences in the fragmented YACs aided the determination of the locus organization.

In general this technique, although a bit tedious, can have multiple applications. For example, it is possible to subclone a part of a gene in two different orientations in this vector. Fragmentations with such constructs will reveal the exact position and orientation of the gene and its exons. Other applications involve the mapping of cDNA selection products (14, 15) and the generation of interstitial deletions (8).

Acknowledgements

I would like to thank Stewart Fabb and Anjum Misbahuddin for providing data, Helen Smith and David Mavkie for the lithium acetate transformation protocol, and the 9th and 11th Wellcome Trust Summer Schools for verifying the protocols.

References

1. Lindsay, S. and Bird, A. (1987). *Nature*, **327**, 335.
2. Bird, A. (1989). *Nucleic Acids Res.*,, **17**, 9485.
3. Bird, A. (1987). *Trends Genet.*, **3**, 342.
4. Larin, Z., Monaco, A. P., and Lehrach, H. (1991). *Proc. Natl. Acad. Sci. USA*, **88**, 4123.
5. Anand, R., Oglivie, D. J., Butler, R., Riley, J. H., Finniear, R. S., Powell, S. J., Smith, J. C., and Markham, A. F. (1991). *Genomics*,, **9**, 124.
6. Hirst, M. C., Rack, K., Nakahori, Y., Roche, A., Bell, M. V., Flynn, G., *et al.* (1991). *Nucleic Acids Res.*, **19**, 3283.
7. Campbell, C., Gulati, R., Nandi, A. K., Floy, K., Hieter, P., and Kucherlapati, R. S. (1991). *Proc. Natl. Acad. Sci. USA*, **88**, 5744.
8. Pavan, W., Hieter, P., and Reeves, R. H. (1990). *Proc. Natl. Acad. Sci. USA*, **87**, 1300.
9. Pavan, W. J., Hieter, P., Sears, D., Burkhoff, A., and Reeves, R. H. (1991). *Gene*, **106**, 125.
10. Lewis, B. C., Shah, N. P., Braun, B. S., and Denny, C. T. (1992). *GATA*, **9**, 86.
11. Ragoussis, J., Senger, G., Trowsdale, J., and Campbell, I. G. (1992). *Genomics* **14**, 423.
12. Burgers, P. M. and Percival, K. (1987). *Anal. Biochem.*, **163**, 391.
13. Gietz, D., St. Jean, A., Woods, R. A., and Schiestl, R. H. (1992). *Nucleic Acids Res.*, **20**, 1425.
14. Hochgeschweder, U. (1992). *Trends Genet.*, **8**, 41.
15. Das Gupta, R., Morrow, B., Marondel, I., Parimoo, S., Goei, V. L., Gruen, J., *et al.* (1993). *Proc. Natl. Acad. Sci. USA*, **90**, 4364.

7

Functional analysis of mammalian genomes using yeast artificial chromosomes

Z. LARIN

1. Introduction

Methods for isolating genes by positional cloning have been enhanced by using yeast artificial chromosomes (YACs) as mapping tools. However, YACs may also be used to isolate genes by functional complementation of a mutant phenotype (1) following transfer of YACs containing the respective region into cultured cells (2, 3) or mice (4, 5, 6). This approach is particularly attractive for isolating murine genes where large regions containing mutant loci have been mapped by genetic markers. Once a YAC is known to encompass a particular gene, further localization of the gene can be obtained by re-introducing a series of YAC deletions, constructed by telomere fragmentation (7) (see Chapter 6), into mammalian cells, and testing for complementation of the mutant phenotype.

Transfer of YACs containing large contiguous regions of DNA into established cell lines allows the correct tissue-specific and site-independent expression of an entire gene by upstream control regions and regulatory elements which may be several to tens of kilobases away from promoter regions and also downstream of the gene. This allows the gene to be studied in its biological context and will be important if the gene is extremely large (e.g. Duchenne muscular dystrophy, 2.5 Mb) (8) or if YACs contain multigenic loci (9).

Transfer of DNA cloned in YACs to mammalian cells or to mice has been achieved by several methods. Conventional methods which transfer DNA to cells such as calcium phosphate precipitation or electroporation may shear or damage large DNA. Instead, YAC DNA has been transferred intact along with the yeast genome by fusing yeast protoplasts with mammalian or embryonic stem cells (5; reviewed in ref. 10) or separated from the yeast genome by pulsed field gel electrophoresis (PFGE) and transferred to cultured cells or mice either by microinjection (6, 11) or lipofection (where DNA/lipid complexes pass into the cell by endocytosis) (3, 4, 12, 13, 14). There are advantages and disadvantages to each of these methods. Protoplast fusion may not impose limitations on the size of DNA transferred, but success has so far been reported only with a limited number of rodent cell lines. The other methods reduce the possibility of contamination from the yeast DNA since YAC DNA is purified by PFGE, but there may be an upper limit to the size of the DNA which can be efficiently transferred intact.

Whichever method is used to transfer YACs to mammalian cells, it is essential to construct a restriction site map of the YAC DNA by PFGE (see Chapter 6) and compare the map to the genomic map (see Chapter 2). It is then important to check whether the DNA has remained intact since exogenous DNA may become rearranged during transfer. Exogenous DNA may also be methylated in mammalian cells and it may be difficult to match the restriction site pattern of DNA obtained from transformants to the YAC DNA using rare-cutting restriction endonucleases. However, the presence of the vector arms can be determined and junction fragments (from the vector arm to the first site in the genomic DNA) can be easily mapped by frequently cutting enzymes. One potentially useful method is RecA-assisted restriction endonuclease (RARE) cleavage (11) which will enable detection of the entire YAC insert in the genomic DNA. In this procedure restriction enzyme sites within the exogenous DNA (e.g. *Eco*RI) are initially protected with homologous oligonucleotides and RecA protein, DNA is then methylated (e.g. with *Eco*RI methylase), and digested with *Eco*RI. Only the sites originally protected from methylation will be sensitive to the enzyme.

2. Modification of YACs prior to transfer to mammalian cells

2.1 Retrofitting YACs in yeast by homologous recombination

Vectors have been constructed which can alter pYAC4 arms by inserting new DNA or replacing the existing arm with different markers by homologous recombination in yeast. This procedure is known as retrofitting. Several different types of vectors are available. Integration vectors can either insert new DNA into the YAC arm or the insert DNA following a single cross-over event at homologous sequences. Fragmentation vectors containing one *Tetrahymena* telomere and a new selectable marker (7) break the YAC by recombining at YAC arm vector sequences or repetitive sequences within the YAC insert. Most vectors contain the bacterial neomycin resistance gene (neo) for selection in mammalian cells and introduce either ADE2, LYS2, or HIS3 as selectable yeast markers (15, 16, 17, 18; Table 1); one vector introduces the thymidine kinase gene into the YAC arm for selection in thymidine-kinase-deficient cells (15) and other vectors recombine at repeat sequences such as

Table 1. Retrofitting plasmids for selection in mammalian cells

Integration vector	Yeast marker	Mammalian selectable marker	Selection in mammalian cells	Reference
(a) Into vector arm	ADE2	Neomycin	G418	18
	LYS2	Neomycin	G418	15, 16, 17
	LYS2	Thymidine kinase	HAT medium	15
(b) Into insert DNA	HIS3	Neomycin	G418	7
	URA3	Neomycin	G418	15
	LYS2	Neomycin	G418	16, 17

SINEs or LINEs within the YAC causing fragmentation (7). YACs can also be modified with conversion vectors which introduce sequence elements for amplification of YACs (19). These vectors increase the copy number of YACs in *Saccharomyces cerevisiae* by introducing a conditional centromere which is inactivated by inducing a Gal 1 promoter, and selecting for expression of the thymidine kinase gene giving multiple copies of the YAC. YACs can be retrofitted with appropriate vector DNA following spheroplast transformation (20) as outlined in *Protocol 1* and Chapter 6, *Protocol 5*.

Protocol 1. Transformation of YACs with retrofitting plasmids

Equipment and reagents

- For many of the reagents and equipment see Chapter 6, *Protocol 5B*
- 1 M sorbitol
- 1 × amino acids: (see Chapter 6, *Protocol 5B*)
- 10 × yeast nitrogen base without amino acids (Difco): 6.7 g/litre dH$_2$O.
- Regeneration medium for transformation plates and top agar: selective medium and 1 M sorbitol
- Selective medium: 2% (w/v) glucose, 1 × amino acids (omitting the appropriate amino acid(s) for selection of the new yeast marker), 0.67% (w/v) yeast nitrogen base (add as 10 × solution after autoclaving). For plates add 2% (w/v) bactoagar (Difco or Oxoid)
- Polyethylene glycol (PEG) solution: 20% (w/v) PEG 6000 (BDH), 10 mM Tris–HCl pH 7.5, 10 mM CaCl
- STC: 1 M sorbitol, 10 mM Tris–HCl pH 7.5, 10 mM CaCl$_2$
- SOS: 25% YPD, 1 M sorbitol, 6 mM CaCl$_2$, and 1 μg of additional amino acid supplements depending on the yeast marker being selected
- TE: 10 mM Tris–HCl pH 7.5, 1 mM EDTA pH 8.0
- Haemocytometer
- Light microscope with phase contrast
- Incubator
- Bench top centrifuge
- 15 and 50 ml plastic tubes (Falcon)

A. *Preparation of protoplasts and vector DNA*

1. Prepare protoplasts as described in Chapter 6, *Protocol 5*. Resuspend protoplasts in STC at a concentration of 4.5 × 10^8/ml. Leave the protoplasts at 18°C until use (not more than 2 h).

2. Linearize the plasmid DNA with the appropriate restriction enzyme and ethanol precipitate. Resuspend in TE at a concentration of 1 μg/μl. Prepare dilutions of the plasmid in TE (e.g. 10 ng/μl, and 100 ng/μl).

Protocol 1. *Continued*

B. *Transformation of yeast protoplasts with the retrofitting vector*

1. Pipette different concentrations of the plasmid (i.e. 10 ng, 100 ng, and 1000 ng) into 15 ml Falcon tubes. Add 150 μl of yeast protoplasts, mix carefully and leave at 18°C for 10 min.

2. Add 1.5 ml of PEG solution and mix carefully by inverting the tube and leave for 10 min at 18°C.

3. Pellet the protoplasts by centrifugation at 300 *g* for 8 min, and then pipette off the supernatant. Resuspend the pellet in 225 μl of SOS medium and leave at 30°C for 30 min. Quickly mix the protoplasts with 5 ml of top agar (kept at 48°C) and pour on to regeneration agar plates lacking the appropriate amino acid for selection of the new yeast marker. Leave at 30°C for 3–4 days for colonies to appear. Streak colonies on to selective medium agar plates lacking the appropriate amino acids for double selection of yeast markers in both YAC arms. Colonies should grow within 1–2 days.

4. Set up YAC minipreps by inoculating 3–5 ml of selective medium lacking sorbitol and appropriate amino acids. To prepare yeast chromosomes in agarose blocks, see Chapter 6, *Protocol 1*.

2.2 Recombination between overlapping YACs

Two or more YACs which contain regions in common can be recombined in yeast to produce a single YAC containing the whole region. YAC clones should contain homologous regions within the insert and be in the correct orientation with respect to each other such that a recombination event will produce only centric fragments. Recombination between YACs can occur between non-chimeric or chimeric YACs at a homologous sequence. Two types of recombination methods have been used, meiotic and mitotic recombination.

Meiotic recombination occurs between YACs introduced into a diploid cell from haploid strains of opposite mating type (i.e. *MAT*a and *MAT*α). Sporulation is induced and analysis of haploid spores by tetrad or random spore analysis and genetic selection is required to obtain recombined YACs containing the appropriate selectable marker and correct phenotype (21, 22). Mitotic recombination is possible by generating isosexual diploids following protoplast fusion of overlapping YAC clones. This procedure also utilizes ultra violet radiation induced recombination during the mitotic cycle and the appropriate YACs can be detected by colour selection (23). Recombinants can also be generated at a reasonable frequency by direct genetic selection of diploids on appropriate selective media containing counter selective agents such as 5-fluoro-orotic acid or α amino-adipate which is toxic to cells containing the parental YAC expressing URA3 or LYS2 respectively (22).

Most YACs generated are in the host strain AB1380 which has mating type *MATa*, and so for meiotic recombination it is first necessary to transfer the YAC to an appropriate haploid strain of opposite mating type *MATα* before recombining with another YAC. This can be done by several different methods. Gel purified YACs can be transferred by protoplast transformation (see Chapter 6, *Protocol 1* for preparation of YAC DNA in agarose, and Chapter 6, *Protocol 5* for transformation of yeast protoplasts), but this method is not suitable for large YACs which do not transfer intact and frequently undergo rearrangements after transformation. A simpler method for maintaining intact YACs involves mating AB1380 containing the YAC with a yeast strain which has a defect in nuclear fusion at the *KAR 1* locus. This generates heterokaryons containing nuclei from both parents which then produce daughter cells or cytoductants containing only one nucleus with cytoplasm from both parents. However, following appropriate selection, rare events involving heterokaryons allow transfer of the YAC from one haploid nucleus to the other (termed YACductants), thus switching the mating type of the nucleus (24, 25). Lastly, YACs can be transferred by crossing AB1380 *MATa* with a strain of opposite mating type *MATα* as described in *Protocol 2* below.

Protocol 2. Recombination of yeast artificial chromosomes

Equipment and reagents

- Sterile distilled water
- YPD agar plates (see Chapter 6, *Protocol 5*)
- Selective medium
- Sporulation medium: 2.5 g/litre yeast extract, 15 g/litre potassium acetate, 0.5 g/litre glucose, 1 × amino acids (see Chapter 6, *Protocol 5*)
- Glusulase (NEN, Dupont; filter, sterilize, and dilute 1:40 in dH₂O)

- 5-fluoro-orotic acid (Sigma)
- α amino-adipate (Sigma)
- Glass beads (Sigma)
- 1.5 ml microcentrifuge tubes
- 15 ml tubes
- Incubator
- Light microscope

A. *Transfer of YAC to opposite mating type strain/generation of diploid cells*

See *Figure 1A*.

1. Streak out the appropriate yeast strains on to YPD or selective medium (for YAC cultures) agar plates. Choose a recipient strain in which it is possible to select for the presence of new auxotrophic markers and complementation of existing markers (e.g. for AB1380 *MATa, ura3, trp1, ade2–1, can1–100, lys2–1, his5* crossed to YS58, *MATα, his4–519, leu2–3; 112, trp1–789, ura3–52*; select for diploids containing the YAC on selective medium lacking uracil, tryptophan, and histidine, see *Figure 1A*). Incubate at 30°C for 2 days.

143

Protocol 2. *Continued*

2. Pick a single colony from each strain and resuspend in 200 μl of dH$_2$O in separate 1.5 ml microcentrifuge tubes.

3. Place two drops of cell suspension from one yeast strain on to a YPD agar plate, and then immediately overlap the area with two drops of cell suspension from the other yeast strain. Allow the area to dry. As a control place separate drops from each parental strain on to the same plate. Incubate the plate at 30°C for 16 h. Check for the formation of diploid cells under the microscope by identifying the formation of dumb-bell-shaped zygotes.

4. Streak out the yeast cells from the mating mixture on to the appropriate selective medium agar plate such that only diploid cells containing the YAC can grow (e.g. for AB1380 and YS58 diploids use selective medium lacking uracil, tryptophan, and histidine, see *Figure 1A*). As a negative control streak out the yeast cells from each parental strain used in the crosses on to the same selective medium agar plates. Incubate the plates for 2–4 days at 30°C.

5. Analyse the YAC content of diploid cells by picking single colonies into 10 ml of selective media (lacking uracil, tryptophan, and histidine), then prepare YAC DNA in agarose as described in Chapter 6, *Protocol 1*. Analyse YAC DNA by PFGE as described in Chapter 6, *Protocol 2*.

B. *Generation of a haploid strain*

See *Figure 1B*.

1. Streak out the appropriate diploid strain containing the YAC on to a YPD agar plate and incubate at 30°C for 2 days.

2. Streak out the yeast cells on to sporulation agar plates and incubate at 30°C for 5–7 days. Check the extent of sporulation by looking at the cells under the microscope.

3. Resuspend some sporulation culture in 200 μl of Glusulase in a 15 ml tube and shake for 1 h at 30°C. Add an equal volume of glass beads and continue shaking for 1 h at 30°C.

4. Add 1 ml of sterile dH$_2$O and vortex for 1 min. Transfer the cell suspension to a 1.5 ml microcentrifuge tube. Pellet the cells by centrifugation at 800 *g* for 1 min and resuspend in 1 ml of dH$_2$O. Plate 1, 10, and 100 μl of cells on to selective medium (e.g. for crosses between AB1380 and YS58 omit uracil and tryptophan) agar plates and incubate at 30°C for 4–5 days. The large colonies that appear are usually diploid cells while the smaller colonies are the haploid spores.

5. Pick several single colonies into 10 ml of selective medium lacking uracil and tryptophan, then prepare YAC DNA in agarose and analyse by PFGE (see Chapter 6, *Protocols 1* and *2*).

6. Determine the mating type of the haploid spores either by crossing them to known *MATa* or *MATα* strains as described in *A* steps 1–4 or by a polymerase chain reaction assay on individual colonies (26).

C. *Meiotic recombination of overlapping YACs*

See *Figure 1C*.

1. Modify one of the vector arms of each YAC by retrofitting with an integrating plasmid as described in Section 2.1, *Protocol 1* (see *Table 1*), to select for recombinant YACs (e.g. ADE2 or LYS2).

2. Then repeat steps 1–4 as described in *A* above using overlapping YACs in strains of opposite mating type *MATa* and *MATα*.

3. Sporulate diploid cells as described in steps 1–5 in *B* above and re-isolate spores on appropriate selective medium. Analyse recombinants by random spore analysis which detects cells containing the recombinant YAC or parental YACs.

D. *Direct selection of mitotic recombinants in diploid cells*

See *Figure 1D*.

1. To select for diploid yeast cells in which YACs have undergone mitotic recombination and produced recombinant YACs, streak out cells directly on to selective medium lacking the appropriate amino acids to select for the new combination of yeast markers in either YAC arm and loss of the parental YACs.

2. In addition, counter-select for yeast colonies expressing the URA3 gene and the LYS2 gene (if these markers were present in the parental YAC arms) by incorporating 5-fluoro-orotic acid or α amino-adipate respectively in the medium which is toxic to the cells.

3. Generate haploid spores containing the recombinant YAC as described in Part *B* (above).

3. Introduction of YAC DNA into mammalian cells

3.1 Lipofection

Lipofection is a simple and efficient method for transferring small DNA molecules to mammalian cells (in transient or stable assays) when compared to existing transfection methods such as electroporation or utilizing mediators such as calcium phosphate or DEAE–dextran. In lipofection, entry of DNA into the cell is mediated by positively charged molecules enveloping negatively charged DNA, producing a net positively charged lipid–DNA complex which

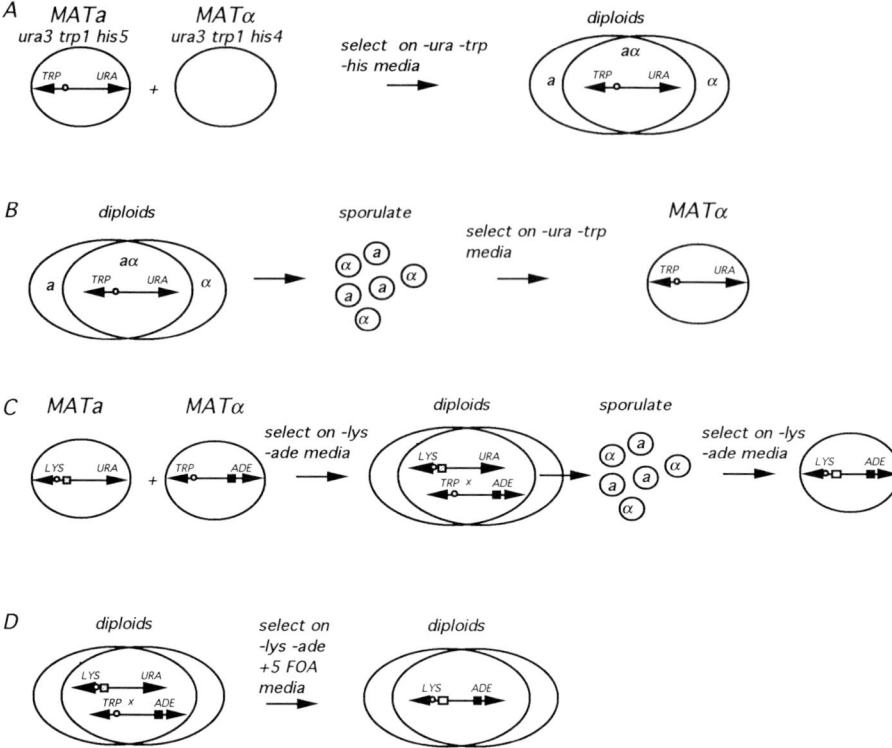

Figure 1. Recombination between overlapping yeast artificial chromosomes. (a) Transfer of a YAC in AB1380 mating type a to a yeast strain (e.g. YS58) of opposite mating type *MATα*. Diploids containing the YAC can be selected on medium lacking histidine (by complementation of the auxotrophic markers *his*5 and *his*4 respectively), uracil, and tryptophan. (b) Diploids containing the YAC are induced to sporulate and haploid spores containing the YAC can be analysed to determine the mating type (*MATa* or *MATα*). (c) Prior to recombination, overlapping YACs in different yeast strains (mating type a or α) are modified by retrofitting either the left or right arm with an integrating vector to introduce new yeast markers, e.g. *LYS*2 and *ADE*2, for selection of the recombinant YAC. Diploids containing both YACs are then selected on medium lacking lysine and adenine and induced to sporulate under minimal conditions. Haploid spores of the correct mating type are then analysed to determine which contain the recombinant YAC. (d) Diploids containing both overlapping YACs can be selected on medium lacking lysine and histidine, but the frequency of diploids containing only the recombinant YAC can be increased by including 5-fluoro-orotic acid in the medium which is toxic to cells expressing *URA3*. This selects against cells containing the parental YAC. Haploid spores containing the recombinant YAC can then be generated from diploids as described in (b).

fuses with negatively charged cultured cells. Since such complexes can efficiently transfer small quantities of plasmid DNA (1 μg) and achieve good transfection in a wide variety of cell types, this method has been used for transferring large DNA as YACs into mammalian cultured cells (3, 14). DNA

Table 2. Liposome reagents

Liposomes	Compound	Reagents
Lipid	Monocationic	DOTMA (Lipofectin, Gibco BRL)
		DOTAP (Boehringer Mannheim)
Lipospermines	Polycationic	DOGS (Transfectam, Promega)
		DPPES (24)
		DOSPA (Lipofectamine, Gibco BRL)

usually integrates into the host genome. YACs containing large genes (human or mouse, 80–400 kb) have also remained essentially intact following transfer by lipofection into embryonic stem cells (4, 12, 13) and the exogenous DNA can subsequently be successfully transmitted through the mouse germ line and expressed appropriately. Different types of lipofection reagents are commercially available (*Table 2*). Monocationic lipids (DOTMA, N-[1-(2, 3-dioleyloxy) propyl]-N,N,N,-trimethylammonium chloride) were initially used in transfection studies, but lipospermines (DOGS, dioctadecylamido-glycylspermine; DPPES, dipalmitoylphosphatidylethanolamylspermine; and DOSPA, 2, 3-dioleyloxy-N-[2 sperminecarboxamido] ethyl-N,N-dimethyl-1-propanaminium trifluoroacetate) are more efficient (7, 8). They can achieve up to 30-fold higher activity than monocationic lipids in many cell types and have been used for transferring large DNA as YACs into mammalian cells.

As the quantity of YAC DNA is the limiting factor, it is necessary to generate concentrated YAC DNA prior to purification from the rest of the host genome by PFGE. One method is to grow a large number of yeast cells and embed them in a small volume of agarose (see *Protocol 3*). Another method involves retrofitting YACs with a vector which will increase the copy number (as described in Section 2.1).

Protocol 3. Preparation of concentrated YAC DNA in agarose

Equipment and reagents

- See Chapter 6, *Protocol 5B* for equipment and reagents
- Novozym: 8 mg/ml (CalBiochem/Nova-Biochem)
- TE$_{50}$: 10 mM Tris–HCl pH 7.5, 50 mM EDTA pH 8.0
- Low melting temperature (LMT) agarose (SeaPlaque GTG, FMC)
- 1 M Dithiothreitol (DTT)

Method

1. Prepare YAC DNA as described in Chapter 6, *Protocol 1* with the following modifications.

2. Inoculate 500 ml of selective medium with 5–10 ml of the overnight culture at 30°C. For 500 ml about 50 agarose blocks can be generated,

---

Protocol 3. *Continued*

    each block containing about 5×10^9 yeast cells per 80 µl of agarose. For 50 blocks resuspend the yeast pellet in 2 ml of SCE so that when mixed with an equal concentration of 1.5% (w/v) LMT agarose the final agarose concentration will be 0.75% (w/v).

3. As an alternative enzyme for generating protoplasts use Novozym to a final concentration of 8 mg/ml and 10 mM DTT in SCE.

4. Leave agarose blocks at 37°C for 4–16 h to spheroplast yeast cells, then wash in lysis solution. Incubate blocks in 50 ml of lysis solution initially for 1 h at 50°C, then change the solution and incubate for a further 24 h at 50°C.

5. Wash the blocks 3–4 times in TE_{50} and then store at 4°C.

3.1.1 Purification of YAC DNA by PFGE

To produce sufficient YAC DNA for each transfection at least 20 concentrated yeast agarose blocks should be purified by PFGE in LMT agarose as outlined in *Protocol 4*. Following electrophoresis, the gel strip containing the YAC should be stored in an appropriate buffer (TE_{50}).

Protocol 4. Isolation of YAC DNA from a LMT agarose gel using PFGE

Equipment and reagents

- TE: 10 mM Tris–HCl pH 7.5, 1 mM EDTA
- TE_{50}: (see *Protocol 3*)
- $0.5 \times$ TBE (0.045 M Tris borate, 0.045 M boric acid, 0.004 M EDTA) or $0.5 \times$ TAE (0.02 M Tris-acetate, 0.0005 M EDTA) for pulsed field gels depending on the apparatus used
- NaCl (5 M stock solution)

- Ethidium bromide (10 mg/ml stock solution)
- Agarase (Sigma), Gelase (Epicenter), or β-agarase I (New England Biolabs)
- LMT agarose (SeaPlaque GTG, FMC)
- PFGE apparatus
- PFGE gel tray and comb
- UV light transilluminator
- Sterile scalpel

Method

1. Prepare a 1% (w/v) LMT agarose gel in $0.5 \times$ TBE or $0.5 \times$ TAE. Tape the teeth of the gel comb (leave at least three teeth on either side without tape for markers). This will form a gel trough which holds at least 12–15 agarose blocks containing YAC DNA. Pour the LMT agarose into the gel tray and allow it to set.

2. Wash the plugs once in TE_{50} and then load into the gel trough. In the outside lanes load either lambda concatamers or yeast markers (see Chapter 2, *Table 2*) and a block containing the appropriate YAC. Overlay the blocks with LMT agarose and leave to set for at least 30 min.

3. Purify YAC DNA by PFGE using appropriate voltage and switching times for adequate separation of the YAC DNA (see Chapter 6, *Protocol 2*).

4. Cut off the outside lanes on either side of the gel trough and stain in 500 ml of buffer containing 30–50 µl of ethidium bromide. Leave the remainder of the gel in TE_{50} at 4°C. Visualize the marker lanes under UV light and mark the position of the YAC in the agarose with a scalpel cut. Reassemble the gel with the marker lanes.

5. Excise the gel slice containing YAC DNA using the marker lanes as a guide. Use a ruler to guide the cutting. Do not cut too wide: the volume of agarose should be about 1–2 ml. Store the gel slice in TE_{50} until use. When preparing for YAC DNA transfer by lipofection into mammalian cells, equilibrate the gel slice in the appropriate buffer recommended by the manufacturer prior to agarase treatment. For example, when using lipospermine (Transfectam, Promega) equilibrate the gel slice in 1 × TE and 100 mM NaCl, then melt 240 ml aliquots at 68°C, cool to 37°C, and add 40 units of agarase/gelase or 50 units of β-agarase I for at least 1 h. Add NaCl to 300 mM final concentration prior to transfection.

3.1.2 Preparation of lipid–DNA complex

In general, the preparation of the lipid–DNA complex for transfection can be carried out according to the manufacturer's instructions. However, since the efficiency of DNA transfection in each cell type can depend on a number of different factors, it may be important to carry out preliminary tests. These may include varying the ratio of DNA:lipid and the time of association prior to transfection, testing the ionic strength of the buffer since DNA in association with polyamines tends to condense unless other cations are present, i.e. Na^+, K^+, Mg^{2+}, and lastly varying the transfection time, usually 6–16 h. Only a small amount of transfection medium is added to the cells during transfection to favour efficient interaction with the cells, and usually fetal calf serum is absent. However, successful transfections can be achieved in the presence of serum but only where the lipid–DNA complex is prepared under serum-free conditions (29). The recommended method for transfection of YAC DNA into mammalian cells by liposomes is outlined in *Protocol 5*.

Protocol 5. Transfection of mammalian cells with liposomes

Equipment and reagents

- Lipofection reagent (see *Table 2*)
- Appropriate tissue culture medium (with and without fetal calf serum) according to cell type
- Tissue culture hood
- Phosphate-buffered saline (PBS): 136 mM NaCl, 2 mM KCl, 10.6 mM Na_2HPO_4, 1.5 mM KH_2PO_4, pH 7.3 (Oxoid)
- Tissue culture incubator
- 6 cm tissue culture dishes (Falcon)

Protocol 5. *Continued*

Method

1. Grow cultured cells to 50–60% confluency in 6 cm dishes. Wash the cells three times in medium without serum and leave in 1 ml of this medium at 37°C.

2. Mix the YAC DNA (isolated as described in *Protocol 4*) with lipofectin reagent. Also mix appropriate plasmid DNA with lipofectin reagent (suggested concentrations 100 ng, 1 μg, and 5 μg to check the transfection efficiency). Some protocols recommend leaving the complex for 30–45 min, while others advise adding the DNA–lipid complex to the cells immediately after mixing. Determine empirically which method gives the best results.

3. Leave the YAC DNA/lipofectin reagent on cells for a minimum of 6–16 h at 37°C. Wash the cells three times with PBS and add 4 ml of medium containing serum.

4. After 48 h add an appropriate concentration of the selective reagent (for example G418 for neomycin resistance). Change the tissue culture medium every 3–4 days and add fresh selective agent.

5. Colonies should appear within 10–14 days. Pick the colonies when they are about 2.5 mm in diameter and expand the cells to at least 10^7 cells for preparing agarose plugs for PFGE (see Chapter 2, *Protocol 1*); also prepare metaphase spreads from 1×10^6/ml cells for fluorescence *in situ* hybridization (FISH) (Section 3.5, *Protocol 8*).

3.2 Transfer of YACs to mammalian cells by yeast spheroplast fusion

DNA has been successfully transferred from one cell type to a different cell type by fusion of heterokaryons with polyethylene glycol (30). This method, described in *Protocol 6*, has been used to transfer YACs to mammalian cells (31, 32), and removes the step of purifying DNA from yeast prior to transfection. DNA is usually transferred in one or a few copies and ordinarily integrates into a single host chromosome relatively intact. Prior to transfer it is necessary to retrofit the YAC with an appropriate selectable marker, i.e. the neomycin resistance gene as described in Section 2.1. In addition, yeast DNA is usually introduced into the recipient cell. While this may cause a change in chromosome appearance at the integrated site (13, 33) it does not appear to prevent the expression of exogenous genes (2) or generation of transgenic mice (5).

Protocol 6. Fusion of yeast protoplasts to mammalian cells by polyethylene glycol

Equipment and reagents

- Reagents for preparing yeast protoplasts as listed in *Protocol 1*
- Reagents for fusing yeast protoplasts to mammalian cells: 50% PEG 1500 (Boehringer Mannheim); to 4 ml add 5 mM CaCl$_2$, 50 μM 2-β mercaptoethanol
- Serum-free tissue culture media
- Appropriate selective agent
- Trypsin
- 15 ml plastic tubes (Falcon)
- 10 cm and 6 cm tissue culture dishes (Falcon)
- 12-well dishes (Falcon)

A. *Preparation of yeast protoplasts*

1. For preparation of yeast protoplasts see *Protocol 1*. Additional note: if YACs are grown in recombination-deficient strains (*RAD* 52 or *RAD* 1) then use a larger inoculum (1/100) for the initial culture as they grow more slowly.

2. Resuspend the cells finally in STC at 1×10^8 cells/ml (i.e. so that the ratio of yeast protoplasts to mammalian cells is approximately 10–50: 1; keep yeast protoplasts at 18°C until use.

B. *Preparation of mammalian cells*

1. Split the mammalian cells 1–2 days before the fusion experiment such that there are about 3×10^6 cells per 10 cm dish. Harvest the cells with trypsin and then pellet by centrifugation at 300 *g* for 5 min and wash the cell pellet three times in serum-free medium. Resuspend the cells at 3×10^6/ml.

2. Pellet 1 ml (1×10^8/ml) of yeast protoplasts by centrifugation at 300 *g* in a 15 ml Falcon tube for 5 min, remove the supernatant then carefully overlay 1 ml of mammalian cells. Pellet the cells by centrifugation again at 300 *g* for 5 min. Carefully remove the supernatant.

3. Gently mix the cell pellet with 50 μl of serum-free tissue culture medium, then add 500 μl of PEG solution. Mix gently by inverting the tube, leave at room temperature for 60–120 sec depending on cell type. Immediately add 5 ml of serum-free medium to the cells (initially dropwise since yeast spheroplasts are fragile), mix gently, and leave for 10 min at room temperature. Pellet the cells by centrifugation at 300 *g* for 5 min and then resuspend the cell pellet in 3–4 ml of tissue culture medium containing serum.

4. Aliquot the cells into 6 cm or 10 cm dishes at an appropriate cell density. After 24 h wash the cells four times with PBS to remove yeast cells, and after 48 h when the cell density should be about 70–80% confluent add the appropriate concentration of selective agent.

Protocol 6. *Continued*

5. Colonies will appear in 10–14 days. Pick the cells when they are about 2 mm in diameter into 12-well dishes containing tissue culture medium plus serum but without selective agent for 24 h then re-apply selection. Expand the cells as described below (see *Protocol 7*) for preparation of DNA for PFGE and metaphase spreads for FISH (see *Protocol 8*).

3.3 Microinjection

An alternative method of transferring large DNA molecules to mammalian cells is by microinjection of DNA into cultured cells (10) or into fertilized mouse oocytes (6). This method also involves purification of YAC DNA by PFGE prior to microinjection. As described above (see *Protocol 4*) this involves excising an agarose slice containing YAC DNA from the gel, melting the agarose at 68°C, and then digesting the agarose with agarase to form a solution containing DNA and digested agarose. However, since partially digested agarose sometimes has difficulty passing through a microinjection needle without clogging, it is necessary to first extract DNA from the agarose solution and reduce the liquid DNA to a suitable working concentration and volume for transfer through a microinjection needle. Extraction of high molecular weight DNA from agarose using phenol can be difficult and causes shearing of large DNA, and methods to extract and concentrate intact YAC DNA from agarose by centrifugation using filter units have been described (10, 12). DNA in agarose is first equilibrated in the presence of either polyamines (0.75 mM spermidine, 0.3 mM spermine) and salt (NaCl > 30 mM), or high salt concentrations alone (100 mM gives the most consistent results) in order to protect the DNA from shear forces, then the agarose is digested prior to centrifugation. YAC DNA purified by PFGE can also be concentrated by allowing electrophoresis of the YAC DNA (placed at 90° to the original PFGE orientation) into high percentage (4%) low melting point agarose by conventional electrophoresis (32). This procedure concentrates the DNA by restricting its mobility. DNA in solution, following digestion of the agarose, can then be dialysed against the appropriate buffer prior to microinjection.

3.4 Analysis of transformants by PFGE

Following introduction of YACs into mammalian cells, transformants can be analysed to detect YAC sequences. Cells can be expanded under selection for preparation of high molecular weight DNA in agarose as described in *Protocol 7*. DNA can then be digested with appropriate restriction enzymes to detect whether YAC DNA has remained intact after introduction into the host genome (see Chapter 2). Following digestion with rare-cutting restriction enzymes, DNA fragments can be separated by pulsed field gel electrophoresis, transferred to nylon filters, which can then be hybridized with the right and left YAC arm specific probes and single copy DNA fragments isolated from the cloned YAC DNA insert in order to detect the correct sized fragments. If YAC DNA was transferred by spheroplast fusion then filters containing DNA can also be probed with the yeast Ty element or total yeast DNA to detect the extent of yeast sequences present in each cell line. DNA can also be analysed by RARE cleavage to detect whether the YAC DNA has remained intact within the host genome (10).

Protocol 7. Preparation of high molecular weight DNA from cultured cells in agarose

Equipment and reagents

- Lysis solution: 0.5 M EDTA pH 8, 1% (w/v) *N*-lauroyl sarcosine, 2 mg/ml Pronase (Boehringer Mannheim)
- TE (see *Protocol 4*)
- PBS (see *Protocol 6*)
- 1.5% (w/v) LMT agarose in PBS
- Perspex block formers (see Chapter 6, *Protocol 1*)
- Restriction enzyme buffer according to manufacturer

Method

1. Harvest the cells and pellet by centrifugation at 500 *g* for 5 min. Wash the cells once in 5 ml of PBS and count them. Pellet the cells by centrifugation at 500 *g* for 5 min and resuspend at a concentration of 2×10^7/ml. Prepare 1.5% (w/v) LMT agarose in PBS and keep at 37°C. Add an equal volume of cells to an equal volume of LMT agarose in PBS, mix, and pipette into taped block formers. Leave to cool on ice for about 20 min.

2. Push the blocks into lysis solution. Leave at 55°C for 24–48 h. Blocks can be stored in this solution at 4°C, or washed in TE and kept at 4°C.

3. Prior to restriction enzyme digests wash the blocks three times in 50 ml of TE and equilibrate in appropriate restriction enzyme buffer on ice (see Chapter 2).

3.5 Detection of YAC sequences in the host genome by fluorescence *in situ* hybridization (FISH)

Prior to transfer into mammalian cells YACs can be analysed quickly by FISH to metaphase chromosome spreads of the appropriate species to determine if the YAC is contiguous or chimeric without the need to end-rescue each arm to check for the presence of the correct sized fragments from the respective chromosome. Following transfection, transformants can also be analysed by FISH (35) as described in *Protocol 8*, by preparing metaphase spreads from each cell line generated. This procedure then detects whether DNA has integrated at single or multiple sites in the host chromosomes and determines the location of the integrated site. It is possible to analyse 24 samples by fluorescence *in situ* hybridization on a single glass microscope slide by compartmentalizing the slide (34). A number of combinations is then possible, either different probes can be hybridized to metaphase chromosomes from one karyotype, or several karyotypes can be analysed simultaneously by the same or different probes. In the largest format up to 96 YAC probes may be hybridized simultaneously to metaphase chromosomes on a glass plate.

Protocol 8. Fluorescence *in situ* hybridization

Equipment and reagents

- PBS (see *Protocol 5*)
- Colcemid: 10 μg/ml stock solution (Gibco)
- Hypotonic solution: 40 mM KCl, 0.5 mM EDTA, 20 mM Hepes pH 7.4
- Fixative solution: methanol:acetic acid, 3:1
- 45% (v/v) acetic acid
- Bio Nick kit (BRL) or Digoxigenin labelling kit (Boehringer Mannheim)
- Cot 1 DNA (BRL): 1 μg/μl
- 20 × SSC: autoclaved stock solution 3.0 M NaCl, 0.3 M Na citrate
- Hybridization buffer: 50% or 65% (v/v) formamide depending on the stringency, 2 × SSC, 10% (w/v) dextran sulphate
- Ribonuclease A (Sigma) 10 mg/ml stock: prepare 100 μg/ml in 2 × SSC
- Formamide (Fluka)
- Rubber glue (Fixo gum, Marabuwerke, GmbH and Co.)
- Phosphate-buffered detergent (PBD): 10 × PBD (Oncor), or 0.1 M phosphate buffer pH 8.0, 0.5% (v/v) non-ionic detergent (nonidet P-40), 0.2% sodium azide

- Ethanol: 70%, 80%, 90%, 100% (v/v)
- Fluorescein-labelled avidin (Oncor or Vector Laboratories)
- Anti-avidin (Oncor or Vector Laboratories)
- Propidium iodide (Sigma)
- Antifade (Vectashield, Vector Laboratories)
- Propidium iodide in antifade (Oncor)
- Microscope slides (BDH): prepare slides by soaking in 0.1 M HCl for 1 h, and then wash several times in 70% (v/v) ethanol and 100% (v/v) ethanol. Leave to air dry.
- 10 cm tissue culture dishes (Falcon)
- 15 ml plastic tubes (Sterilin)
- Tissue culture hood
- Tissue culture incubator
- 1.5 ml microcentrifuge tubes
- Microcentrifuge
- Speed vacuum centrifuge
- Plastic coverslips (Oncor)
- Tweezers (Millipore)
- Fluorescence microscope

A. *Preparation of metaphase spreads from cultured cells*

1. Grow the cells to about 70% confluency in a 10 cm dish. Add 300 μl of colcemid directly to the medium and leave for 75 min at 37°C.

2. Wash the cells twice in PBS, and add 5 ml of pre-warmed (37°C) hypotonic solution. Leave at 37°C for 25 min exactly, then pellet the cells by centrifugation in a 15 ml tube (Sterilin) at 500 *g* for 5 min.

3. Resuspend in a few drops of ice-cold fixative and mix gently until the solution becomes cloudy. Then slowly and dropwise, mixing after each drop, add about 1 ml of fixative solution. Keep on ice for 20 min then top up slowly (while mixing) with about 12 ml of ice-cold fixative and leave at −20°C overnight.

4. Pellet the cells by centrifugation at 300 *g* for 5 min and wash in 10 ml of fresh, cold fixative solution. Repeat three times, and finally resuspend the cells in about 1 ml of fixative solution. Drop the cells on to clean glass microscope slides directly or drop 45% (v/v) acetic acid in

dH$_2$O first on to the slide to help disperse chromosomes. Leave the slides to air dry then view under the microscope. If the cells are too dense or too sparse, spin the cells again at 500 *g* for 5 min and resuspend in fixative solution at an appropriate concentration.

B. *Preparation of YAC DNA probe*

1. Prepare biotinylated or digoxygenin-labelled YAC DNA according to the manufacturer's instructions. Label 1 µg of total yeast DNA containing the YAC and use 200 ng of DNA per slide.

2. Aliquot labelled YAC DNA into a 1.5 ml microcentrifuge tube and add Cot 1 DNA (4 µg) for competition of repetitive sequences. Dry down the DNA in a speed vacuum centrifuge and resuspend in hybridization buffer. Leave at 37°C for 1 h to solubilize the DNA. Denature DNA at 90°C, chill briefly on ice, spin, and pre-anneal DNA at 37°C for 20 min before applying to the slide.

C. *Preparation of metaphase chromosomes for FISH*

1. Use metaphase chromosomes which were spread at least the day before to prevent over-denaturation of chromosomes. Alternatively, 'age' slides in 2 × SSC at 37°C for at least 2 h before use. Mark the edge of the slide with pencil to indicate the area where most metaphases are located.

2. Treat the slides with ribonuclease A, 100 µg/ml in 2 × SSC, for 30 min at 37°C. Dehydrate the slides sequentially in 70%, 80%, 90%, and 100% (v/v) ethanol.

3. Denature the chromosomes in 70% (v/v) formamide in 2 × SSC at 70°C for 2 min. For each additional slide raise the temperature by 1°C. It is better not to denature more than three slides at a time to prevent over-denaturation of chromosomes.

4. Immediately dip the slides into ice-cold 70% (v/v) ethanol for 2 min and then sequentially in cold 80%, 90%, and 100% (v/v) ethanol. Leave to air dry.

5. Add the probe to the slide in the marked area; draw liquid out in a line, then carefully lever down the coverslip to avoid air bubbles.

6. Seal the slide with rubber glue then leave at 37°C in a sandwich box with tissue saturated with 50% (v/v) formamide in 2 × SSC overnight.

D. *Detection of YAC sequences*

1. Remove rubber cement carefully with tweezers without dislodging the coverslip.

2. Soak the slide in 4 × SSC. Usually the coverslip eases off naturally in solution or can be flicked off gently.

Protocol 8. *Continued*

3. Wash the slide in two changes of 50% (v/v) formamide, 2 × SSC at 42°C for 20 min, then two changes of 2 × SSC for 15 min at 37°C. Rinse the slide once in cold 2 × SSC, and then finally in PBD.

4. Add 60 µl of fluorescein-labelled avidin to the slide and seal with a plastic coverslip. Leave at 37°C for 20 min in a box with tissue saturated in 2 × SSC.

5. Wash the slide 4–5 times, 2 min each in PBD. Apply 60 µl of anti-avidin. Leave at 37°C. Wash the slide 4–5 times, 2 min each in PBD, then apply 60 µl flourescein-labelled avidin for 20 min. Wash slides 4–5 times, 2 min each in PBD. Apply about 20 µl of propidium iodide in antifade to the slide to stain chromosomes and view under UV light by fluorescence microscopy.

Acknowledgements

I thank Chris Tyler-Smith and Stephen Taylor for helpful comments on the chapter, and Stephen Taylor for *Protocol 2*. ZL was supported by the Medical Research Council and the Cancer Research Campaign.

References

1. Larin, Z. and Lehrach, H. (1990). *Genet. Res. Camb.*, **56**, 203.
2. Huxley, C., Hagino, Y., Schelssinger, D., and Olson, M. V. (1991). *Genomics*, **9**, 742.
3. Strauss, W. M. and Jaenisch, R. (1992). *EMBO J.*, **11**, 417.
4. Strauss, W. M., Dausman Beard, C., Johnson, C., Lawrence, J. B., and Jaenisch, R. (1993). *Science*, **259**, 1904.
5. Jakobovits, A., Moore, A. L., Green, L. L., Vergara, G. J., Maynard-Currie, C. E., Austin, H. A., and Klapholz, S. (1993). *Nature*, **362**, 255.
6. Schedl, A., Montoliu, L., Kelsey, G., and Schutz, G. (1993). *Nature*, **362**, 258.
7. Reeves, R. H., Pavan, W. J., and Hieter, P. (1992). In *Methods in enzymology* (ed. R. Wu), Vol. 216, pp. 584–603. Academic Press, London.
8. Monaco, A. P., Walker, A. P., Millwood, I., Larin, Z., and Lehrach H. (1992). *Genomics*, **12**, 465.
9. Peterson, K. R., Zitnik, G., Huxley, C., Lowrey, C. H., Gnirke, A., Leppig, K. A., Papayannopolou, T., and Stamatoyannopoulos, G. (1993). *Proc. Natl. Acad. Sci. USA*, **90**, 11207.
10. Huxley, C. and Gnirke, A. (1991). *BioEssays*, **13**, 545.
11. Gnirke, A., Huxley, C., Peterson, K., and Olson, M. (1993). *Genomics*, **15**, 659.
12. Choi, T. K., Hollenbach, P. W., Pearson, B. E., Ueda, R. M., Weddell, G. N., Kurahara, C. G., Woodhouse, C. S., Kay, R. M., and Loring, J. F. (1993). *Nature Genet.*, **4**, 117.

13. Lamb, B. T., Sisodia, S. S., Lawler, A. M., Slunt, H. H., Kitt, C. A., Kearns, W. G., Pearson, P. L., Price, D. L., and Gearhart, J. D. (1993). *Nature Genet.*, **5**, 22.
14. Larin, Z., Fricker, M. D., and Tyler-Smith, C. (1994). *Hum. Mol. Genet.*, **3**, 689.
15. Srivastava, A. K. and Schlessinger, D. (1991). *Gene*, **103**, 53.
16. Riley, J. H., Morten, J. E. N., and Anand, R. (1992). *Nucleic Acids Res.*, **20**, 2971.
17. Davies, N. P., Rosewell, I. R., and Brüggemann, M. (1992). *Nucleic Acids Res.*, **20**, 2693.
18. Markie, D., Ragoussis, J., Senger, G., Rowan, A., Sansom, D., Trowsdale, J., Sheer, D., and Bodmer, W. F. (1993). *Somat. Cell Mol. Genet.*, **19**, 161.
19. Smith, D. R. Smyth, A. P., Strauss, W. M., and Moir, D. T. (1993). *Mamm. Genome*, **4**, 141.
20. Burgers, P. M. J. and Percival, K. J. (1987). *Anal. Biochem.*, **163**, 391.
21. Den Dunnen, J. T., Steensma, H. Y., Grootscholten, P. M., and van Ommen, G. J. B. (1992). In *Techniques for the analysis of complex genomes* (ed. R. Anand), pp. 197–215. Academic Press, London.
22. Rotomondo, F. and Carle, G. (1994). *Nucleic Acids Res.*, **22**, 1208.
23. Ragoussis, J., Trowsdale, J., and Markie, D. (1992). *Nucleic Acids Res.*, **20**, 3135.
24. Hugerat, Y., Spencer, F., Zenvirth, D., and Simchen, G. (1994). *Genomics*, **22**, 108.
25. Spencer, F., Hugerat, Y., Simchen, G., Hurko, O., Connelly, C., and Hieter, P. (1994). *Genomics*, **22**, 118.
26. Huxley, C., Gren, E. D., and Dunham, I. (1990). *Trends Genet.*, **6**, 236
27. Ciccarone, V., Hawley-Nelson, P., and Jessee, J. (1993). In *Focus* (ed. D. Cupo), Vol. 15, pp. 73–80. Life Technologies, USA.
28. Loeffler, J.-P. and Behr, J.-P. (1993). In *Methods in enzymology* (ed. R. Wu), Vol. 217, pp. 599–618. Academic Press, London.
29. Hawley-Nelson, P., Ciccarone, V., Gebeyehu, G., and Jessee, J. (1993). In *Focus* (ed. D. Cupo), Vol. 15, pp 80–3. Life Technologies, USA.
30. Ward, M., Scott, R. J., Davey, M. R., Clothier, R. H., Cocking, E. C., and Balls, M. (1986). *Somat. Cell Mol. Genet.*, **12**, 101.
31. Pavan, W. J., Heiter, P., and Reeves, R. H. (1990). *Mol. Cell Biol.*, **10**, 4163.
32. Pachnis, V., Pevny, L., Rothstein, R., and Constantini, F. (1990). *Proc. Natl. Acad. Sci. USA*, **87**, 5109.
33. McManus, J., Perry, P., Summer, A. T., Wright, D. M., Thompson, E. J., Allshire, R. C., Hastie, N. D., and Bickmore, W. A. (1994). *J. Cell Sci.*, **107**, 469.
34. Schedl, A., Larin, Z., Montoliu, L., Thies, E., Kelsey, G., Lehrach, H., and Schutz, G. (1993). *Nucleic Acids Res.*, **21**, 4783.
35. Trask, B. J. (1991). *Methods Cell Biol.*, **35**, 4.
36. Larin, Z., Fricker, M. D., Maher, E., Ishikawa-Brush, Y., and Southern, E. M. (1994). *Nucleic Acids Res.*, **22**, 3689.

8

PFGE in the study of a bacterial pathogen (*Haemophilus influenzae*)

D. W. HOOD

1. Introduction

Bacterial chromosomes are relatively small and much simpler to analyse than the larger, more complex, genomes of higher eukaryotic organisms. Pulsed field gel electrophoresis (PFGE) is capable of resolving DNA molecules up to ten million nucleotides in length which should enable one to view the complete genome of most bacteria as a single electrophoretic band. Restriction endonuclease digestion of bacterial genomic DNA and resolution of the resulting fragments by PFGE has facilitated the construction of physical maps of bacterial chromosomes. Since the first bacterial chromosomal restriction map was determined for *Escherichia coli* (1) these have now been obtained for an ever-increasing number of bacterial species. They are proving to be extremely useful in studying many aspects of the genome and genetic and molecular studies on chromosome organization and function can now be carried out directly on intact chromosomes. In particular, for the subject of this chapter, genomic maps have enhanced our investigation of the genetic organization of pathogenicity determinants in the bacterium *Haemophilus influenzae*. These studies will serve to illustrate various aspects and uses of pulsed field gel electrophoresis relevant to the analysis of bacterial chromosomes.

2. Physical map of a bacterial pathogen

Over the past ten years there has been a significant increase in research in the area of bacterial pathogenicity. New bacterial pathogens have been confirmed but it is more the advances in molecular biology and nucleic acid technology which have allowed rapid advances in the study of many bacterial pathogens. Ubiquitous techniques such as transposon mutagenesis have allowed us to create mutants in many poorly understood species. Historically, the study of a wide range of bacteria, including *Haemophilus influenzae*, has been impeded by a lack of reliable genetic maps. Partial genetic maps have been constructed in bacteria by co-transformation experiments, conjugational disruption, or, if suitable phage are available, by co-transduction frequencies. These methods are dependent on having suitable numbers of selectable markers and are often severely limited by the maximum distance which can be studied in any one experiment. Many *H. influenzae* strains have relatively efficient inducible-DNA uptake and transformation systems but these have only proved efficient for mapping closely-linked markers. Co-transformation frequencies vary with the size of fragment and the distribution of specific uptake sequences within the transforming DNA.

Mapping the positions of genes in *H. influenzae* has been impeded by a lack of natural conjugative or transducing systems which have proved so invaluable in genetic linkage studies in other bacteria. This is accentuated by the fastidious nature of the microorganism. *H. influenzae* has an absolute requirement for exogenous iron (haem) and NAD and only grows well on a complex nutrient support such as brain, heart infusion (BHI) medium. Thus nutritional mutants, which have proved essential for the construction of the comprehensive genetic maps of *E. coli* and *Salmonella typhimurium*, cannot be easily isolated and characterized.

2.1 A physical map of *H. influenzae*

Encapsulated *H. influenzae* type b strains are a major cause of meningitis and other life-threatening infections in childhood. The number of genes from these strains which have been identified, cloned, and characterized has increased significantly in recent years. Some of these genes are associated with the biosynthesis and localization of two of the major cell-surface structures, the exopolysaccharide capsule and lipopolysaccharide (LPS). These are known to be two of the main virulence determinants in this organism (2, 3). As the number of characterized genes grew larger it became increasingly important to map their genetic positions. A physical map of the *H. influenzae* genome was required and as a result, the second ever bacterial genomic map to be constructed was for the type d strain, Rd (4). This was the first map derived solely after restriction analysis unlike the *E. coli* map which was based on an extensive background of genetic linkage data.

The organization of genes for pathogenic determinants has often been shown to be an important factor in their function. It is thought that grouping of virulence-related genes may be important for control and in the potential transfer of pathogenic functions between related strains and to other species.

Physical maps of the genome from several strains of *H. influenzae* are now available (4, 5). In our laboratory we have concentrated our efforts on a representative type b strain, Eagan, which was isolated from the cerebrospinal fluid of a child with meningitis (6) and has since been used extensively in genetic and virulence studies (7).

3. Pulsed field gel electrophoresis

Standard laboratory procedures for bacterial chromosomal DNA preparation yield relatively low molecular weight, randomly sheared DNA molecules. Of singular importance for PFGE analysis is the isolation and maintenance of intact high molecular weight chromosomal DNA. This is best obtained from actively growing bacterial cells harvested at the mid-log phase of growth. Bacterial cells are then embedded in agarose blocks and chromosomal DNA is prepared *in situ* by lysing the bacterial cells then removing the envelope and other debris and inactivating potentially harmful nucleases and proteases. Intact chromosomal DNA is then digested *in situ* prior to electrophoresis.

3.1 Preparation of bacterial chromosomal DNA

The procedures described in *Protocol 1* will be useful for preparing chromosomal DNA from many Gram-negative bacteria and should require modification only in the lysis procedure for other bacterial species.

Protocol 1. Bacterial culture and preparation of chromosomal DNA

Equipment and reagents

- Shaking incubator at 37°C
- Brain, heart infusion broth (BHI) (Oxoid, Unipath Ltd)
- Nicotinamide adenine dinucleotide (NAD), 1mg/ml stored at −20°C
- Haemin, 1 mg/ml stored at 4°C
- Chloramphenicol (in 96% ethanol), 10 mg/ml
- Spectrophotometer
- Refrigerated centrifuge set at 4°C

- Wash buffer: for *H. influenzae* we use solution 21 (8) containing salts and amino acids. With less fastidious organisms any general wash/maintenance buffer routinely used for that organism after cell harvesting can be substituted. A general type buffer would be 100 mM NaCl, 20 mM Tris, 20 mM EDTA, pH 7.6
- Low melting temperature (LMT) agarose
- Perspex block mould (e.g. with dimensions 15 × 15 × 3 mm)

Method

1. Grow 200 ml cultures of bacteria in liquid medium. For *H. influenzae* strains we use brain, heart infusion broth (BHI) supplemented with NAD (2 μg/ml) and haemin (10 μg/ml) at 37°C with shaking. Cultures were inoculated from a fresh plate and the growth followed to an OD_{490nm} of 0.4–0.5 (approximately 2×10^9 colony forming units/ml).

2. Add chloramphenicol to 20 μg/ml and continue incubation at 37°C for 1 h.

3. Chill culture on ice for 20 min then harvest bacteria at 3000 *g*, 4°C.

4. Resuspend cells in 20 ml of wash buffer pre-chilled on ice.

5. Pellet cells at 3000 *g*, 4°C, then resuspend in 3 ml of cold wash buffer.

D. W. Hood

Protocol 1. *Continued*

6. Warm to 37°C then add 1 ml of 3% (w/v) low melting temperature agarose in wash buffer at 37°C.

7. Maintain bacteria/agarose mixture at 37°C and pipette 0.2 ml aliquots into a Perspex block mould.[a] Should a mould not be available then the mixture can be solidified in small bore silicon tubing using a syringe for loading and release.

8. Solidify at room temperature then harden at 4°C for 15 min.

[a] 0.2 ml is suitable for a block former with dimensions 15 × 15 × 3 mm.

In our experience, bacteria cultured to higher cell densities will yield greater quantities of chromosomal DNA but will give increased non-specific background after subsequent electrophoresis. Bacteria/agarose blocks should not be stored for extended periods without further treatment.

In the cell lysis procedure described in *Protocol 2* only high grade reagents should be used. This treatment solubilizes all of the components of the bacterial cell other than the DNA. Non-DNA components are then removed by extensive washing of the agarose blocks.

Protocol 2. Treatment of bacteria/agarose blocks

Equipment and reagents

- Bacterial cells prepared in agarose blocks (see *Protocol 1*)
- NDS buffer: 0.5 M EDTA, 10 mM Tris pH 9.5, 1% (w/v) sodium-*N*-lauryl sarcosine[a]
- Proteinase K (Sigma)
- TC buffer: 10 mM Tris pH 7.5, 1 mM CDTA[b]
- Phenylmethylsulfonyl fluoride (PMSF), 100 mM stock solution stored at −20°C[c]
- 0.5 M EDTA pH 8.0

Method

1. Carefully remove agarose blocks from mould.

2. Wash batches of plugs in 40 ml NDS buffer containing Proteinase K (2 mg/ml) for 1 h at 37°C then 24 h at 50°C.

3. Replace buffer and incubate blocks at 50°C for a further 24 h.

4. Rinse blocks in NDS buffer alone at 50°C for 2 h to remove Proteinase K.[d]

5. Rinse blocks in two 40 ml lots of TC buffer with 2 mM phenylmethylsulfonyl fluoride[d] (PMSF) at room temperature.

6. Incubate blocks in 40 ml TC buffer/PMSF at 50°C for 30 min.

7. Repeat step 6.

8. Rinse blocks in three 40 ml lots of TC buffer to remove PMSF.

9. Store blocks in 0.5 M EDTA (pH 8.0) at 4°C.

---

[a] Sodium lauryl sulphate (SLS) can also be used.
[b] CDTA is 1,2-cyclohexylenedinitrilotetraacetic acid; EDTA can be substituted.
[c] **Caution:** PMSF is extremely hazardous to the respiratory tract. Dispense only in well ventilated areas. PMSF is inactivated by temperature and high pH. Aqueous solutions can be discarded after several hours at room temperature under alkaline conditions (pH > 8.5).
[d] Blocks can be stored for short periods after this step if required.

Blocks can be stored at 4°C for up to six months but have been used after 12 months storage with little loss of DNA quality.

3.2 Digestion of DNA/agarose blocks

Cut portions from the stored blocks to provide the required amount of DNA for digestion. A 200 ml bacterial culture gives 19–20 blocks by this method, each containing 10–20 μg of chromosomal DNA.

When calculating the amount of DNA required for digestion for each particular application of PFGE, take into account the dimensions of the gel and the slot former, and the estimated size and number of digested DNA bands. Typically, use 4–6 μg per digestion and run 0.5–3.0 μg per gel lane in the PFGE system described in Section 3.3.

3.2.1 Choosing restriction enzymes

The choice of restriction enzymes must be made for each organism. For efficient initial mapping of bacterial chromosomes it is advisable to use enzymes which have relatively few sites and give larger fragments from the target DNA. In general, only restriction enzymes with six base-pair recognition sites, or greater, are useful. The primary screening for suitable enzymes can be carried out using high molecular weight chromosomal DNA purified by conventional methods then digested and fractionated by standard agarose gel electrophoresis. Any enzyme producing a large number of small fragments (< 10 kb) is unlikely to be useful for PFGE and mapping. A major factor in selecting suitable restriction enzymes is the base composition (%G+C content) of the target DNA. For example, in the case of *H. influenzae*, with a low %G+C content of 37%, enzymes which have proved useful are those with G/C biased recognition sequences such as *Sma*I (CCCGGG), *Apa*I (GGGCCC), *Not*I (GCGGCCGC), and *Eag*I (CGGCCG). Data from sequenced genes show that runs of C and G bases are rare in *Haemophilus* DNA and indeed *Not*I cuts at only a single site in the entire 2100 kb of the *H. influenzae* strain Eagan chromosome (5). Conversely, physical maps of bacteria with genomic DNAs of high %G+C content (e.g. *Streptomyces*, 73%) have been made with enzymes with A/T base rich recognition sites such as *Dra*I (AAATTT) and *Ase*I (ATTAAT) (9). For organisms with DNAs of intermediate %G+C contents, enzymes with A/T- or G/C-rich recognition sites should be of use. Base composition of DNA has no significant effect on mobility under normal PFGE conditions. Once selected, these enzymes can be used to digest the bacterial DNA fixed in agarose blocks.

Protocol 3. Digestion of DNA/agarose blocks

Equipment and reagents

- Chromosomal DNA prepared in agarose blocks (see *Protocol 2*)
- TC buffer (*Protocol 2*)
- Microcentrifuge tubes
- Restriction enzymes and buffers (made according to the manufacturers' instructions)

Method

1. Wash blocks, or portions of blocks, in 0.5 ml TC buffer in microcentrifuge tubes for 30 min at room temperature.

2. Repeat step 1.

3. Transfer blocks to fresh microcentrifuge tubes containing restriction enzyme buffers for the relevant enzymes. Incubate at 40°C for 30 min.

4. Transfer blocks to fresh restriction enzyme buffer then incubate at room temperature for a further hour.

5. Transfer blocks to microcentrifuge tubes containing 150 μl of restriction enzyme buffer and heat at 65°C for 15 min to melt the agarose.[a]

6. Cool to 37°C for 5 min, add the restriction enzyme[b] then incubate at the temperature suitable for optimum enzyme activity.[c] Incubate for 16–20 h.

7. Add a further aliquot of enzyme[b,c] and continue incubation for a further 6–8 h.

8. Heat digested DNA at 65°C for 5 min to inactivate the enzyme then use immediately or store at 4°C until required.

---

[a] DNA in blocks can be digested without melting but for some enzymes we find that digestion is reproducibly more complete after melting.
[b] The minimum amount of restriction enzyme required varies but for most common enzymes we find that complete digestion is obtained with about 20 units of enzyme added over the two aliquots.
[c] For enzymes incubated at 37°C or above the agarose should stay molten. For enzymes incubated at lower temperatures the agarose will set and must be remelted before the second aliquot of enzyme is added.

Other factors to take into account when selecting restriction enzymes are their relative stabilities and any potential methylation/modification of the target DNA. The purity of the enzyme preparation itself is also important.

High-purity genome-grade enzymes are becoming increasingly available from the manufacturers.

3.3 Pulsed field gel electrophoresis of digested DNA

Following restriction enzyme digestion, DNA must be fractionated through agarose gels to separate the fragments as described in *Protocol 4*. The theory of PFGE is discussed in detail in Chapter 1 of this book so only a brief summary of some relevant points will be given here.

During conventional electrophoresis, negatively charged DNA molecules move through the agarose (a molecular sieve) under a constant forcing electric field, at a rate dependent on their size. Large molecules (> 40 kb) cannot be separated efficiently because their size approaches that of the molecular sieve and other factors influenced by their structure and kinetics become limiting. However, if DNA molecules are subject to alternating electric fields (pulsed fields) at angles other than the normal moving force then the folding of the DNA molecules is changed and they realign. Larger DNA molecules take longer to realign in such a way that they can be resolved in repeating pulsed fields. Effective separation of any particular size range of DNA fragments depends on the apparatus and the precise conditions used. The amount of DNA required for any purpose will also vary with the gel system being used.

Of the several modifications of PFGE apparatus available, we have used a contour-clamped homogeneous electric field (CHEF) electrophoresis system (10). Multiple electrodes arranged in a hexagon around the gel generate the electric field at angles of 60° or 120° depending on the polarity. This gives a uniformly clamped field at all points of a gel. Straight bands and lanes are more readily obtained with this system than some others and DNA molecules over 10 Mb can be resolved.

Protocol 4. Gel and electrophoresis conditions

Equipment and reagents

- 1.5% (w/v) and 0.8% (w/v) high melting temperature agarose (Sigma) in 0.5 × TBE (0.045 M Tris, 0.045 M boric acid, 1 mM EDTA, pH 8.3)
- PFGE apparatus and gel tray (15 cm × 15 cm)
- Digested chromosomal DNA in agarose (see *Protocol 3*)
- DNA size standards (λ oligomers) (Boehringer Mannheim) [a]
- Ethidium bromide (1 μg/ml) [b]

Method

1. Prepare gel. These are typically 15 × 15 cm made from 120 ml of 1.5% (w/v) high melting temperature agarose in 0.5 × TBE. We prefer 1.5% agarose for gels to keep the DNA bands tight. Melt the agarose to homogeneity. If open-necked flasks are used then compensate for

---

165

Protocol 4. *Continued*

weight loss by evaporation to maximize reproducibility between repeat gels. Take care to use a level surface. Allow gel to set. For preparative purposes, low melting temperature agarose can be substituted.

2. Heat the digested DNAs at 65 °C for 10 min to melt agarose prior to electrophoresis. If digested in blocks, insert the agarose directly into the wells without melting.

3. Remove the gel comb and pipette the samples into the dry wells of the gel. Include suitable molecular weight standards, e.g. λ oligomers. If supplied as agarose blocks these can be melted or slices can be inserted directly into the wells.

4. If samples do not fill the wells, or wells are left empty, fill them flush to the gel surface with 0.8% (w/v) agarose in 0.5 × TBE. Allow to set.

5. Position the gel in the electrophoresis apparatus filled with running buffer (0.5 × TBE) to at least 2 mm above the gel surface. Programme running parameters.^{c} Gels are generally run cool (4 °C–14 °C) to dissipate heat and maintain band sharpness.

6. After electrophoresis, stain gels with ethidium bromide (1 μg/ml) for 15 min. Photograph gel if required. To visualize any minor DNA species, destain the gel in water for 30 min prior to photography. DNA bands with as little as 1 ng of DNA can be detected.

---

^{a} λ oligomers can be made by ligation of intact λ DNA (11).
^{b} **Caution**: ethidium bromide is a potent mutagen and is toxic.
^{c} Typically, gels were run at 150 V for 24 h with an initial pulse time of 10 sec and final pulse time of 40 sec in the BioRad CHEF-DRII system. Migration rates of DNA molecules are dependent on pulse times, voltage (field strength), pulse angle, and run time. The exact electrophoretic conditions required will vary with the apparatus used and the size of fragments to be separated.

In general it is better to start a mapping project with gel electrophoresis parameters to give as broad a range of separation as possible, then to alter them to focus on areas of particular interest. The conditions selected must be a compromise to allow effective separation of the highest molecular weight species but not to allow any useful low molecular weight species to run off the bottom of the gel. Most often, optimal resolution is found with reorientation times comparable to the pulse time. The effect of increasing the pulse time is to widen the range of molecular size separation with a concomitant reduction in the separation between bands. To achieve a compromise between the overall separation and interband resolution, two or more pulse times can be used during a single run. Size markers are very useful for monitoring the resolution of gels when unknown digests are being used.

Care must be taken not to overload the samples as increasing amounts of DNA cause gel bands to be retarded and over-estimated in size. Closely running bands will also merge to mask their true molecular weights. Computer programs are available which allow the operator to predict the correct parameters for any given situation. Information on running parameters are normally supplied by the manufacturers of the pulsed field apparatus.

4. Construction of a genomic restriction map

4.1 Analysis of pulsed field gels

Figure 1 shows a typical pulsed field gel run with *H. influenzae* chromosomal DNA cut with a variety of restriction enzymes. Electrophoretic conditions were chosen to resolve optimally fragments greater than 20 kb. As stated earlier, the choice of enzymes is of crucial importance to the subsequent analysis of data. For the digests shown in *Figure 1*, *Rsa*I produces four, *Sma*I 15, *Eag*I 31, and *Nae*I 35 DNA species (5). The non-uniform intensities indicate that some bands constitute up to four individual DNA species. None of these enzymes would show any discernible digestion of chromosomal DNA under standard electrophoretic conditions. Not all G/C base pair biased enzymes tested on our DNA gave good results. Digestion with *Apa*I (GGGCCC) produced variable banding patterns, *Sac*II (CCGCGG) produced too many bands, and *Nar*I (GGCGCC) did not appear to cleave *H. influenzae* DNA. *Not*I (GCGGCCGC) is a particularly useful enzyme as it cuts the *H. influenzae* strain Eagan DNA only once and the recognition site contains within it an *Eag*I site (CGGCCG). This helped as a reference point to line up the maps relative to one another during subsequent analysis.

Samples which have been degraded by nucleases during preparation will give agarose blocks with low yields of DNA and will show small DNA species when subject to electrophoresis directly after preparation. Batches of new blocks can be tested prior to experiment. Problems can also be encountered with nonspecific degradation of DNA during enzyme digestion and should this occur for several enzymes, or after fresh batches of enzyme/buffer have been tried, then it is best to discard the agarose blocks and remake the chromosomal DNA.

To construct a genomic map one must first interpret the gels and determine the size of DNA bands obtained for each digest. Accomplish this simply by measuring the distance travelled by each band and determining the size against a calibration curve constructed from the λ oligomer or other standards. Allow some tolerance in the sizes determined by experiment. This varies with the system and the conditions used but as a general rule allow a 5% error for initial sizing data. Once obtained, display the information as an idealized fractionation pattern as shown in *Figure 2*. Such results are based on the data from several independent DNA digestions and gels. Confirmation that the complete banding pattern has been identified and that multiple bands have been correctly resolved should come after the average genome size has been calculated for each enzyme digest.

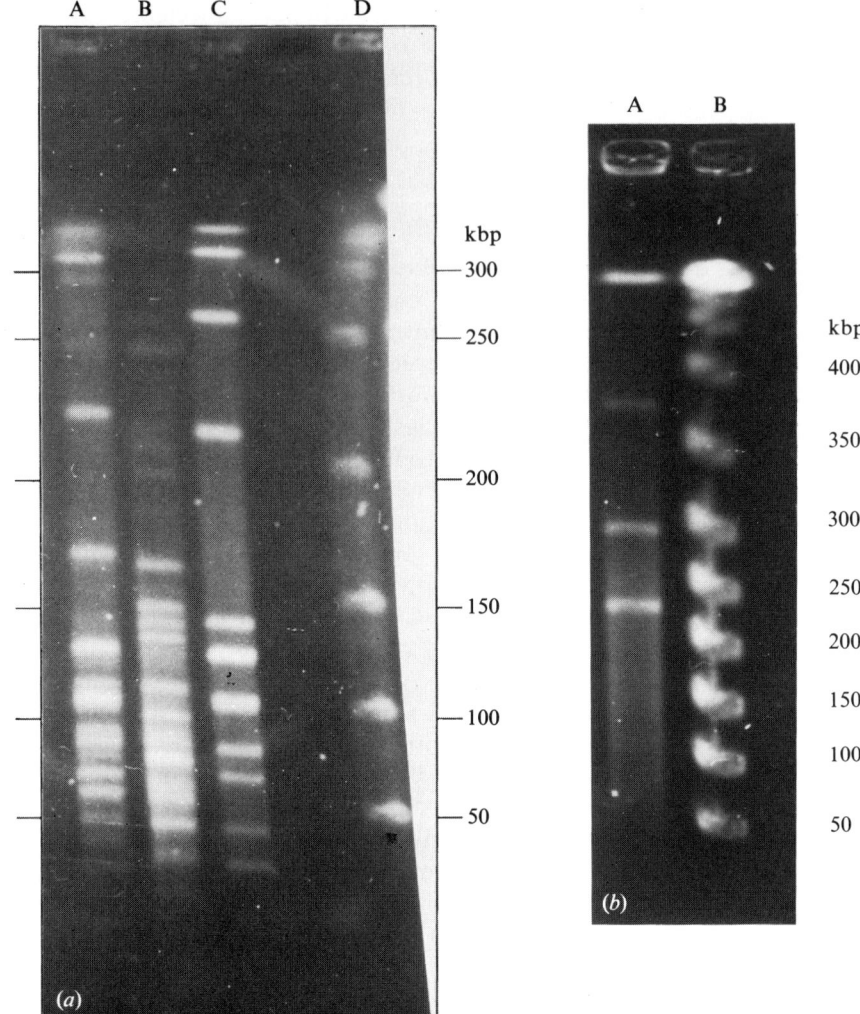

Figure 1. CHEF gel electrophoresis of DNA of *H. influenzae* strain Eagan digested with (a) *Eag*I (A), *Nae*I (B), or *Sma*I (C) and (b) *Rsr*II (A). DNA digestion and electrophoresis conditions were as described in the text with 0.5–2.5 μg of DNA loaded per gel lane. Bacteriophage λ DNA oligomers ((a) D and (b) B) were used as molecular weight markers. The gel was photographed after staining with ethidium bromide.

4.2 Interpretation of data

Analyse and compile the gel information to obtain a meaningful physical map. This is relatively easy with restriction enzymes which produce only several bands as DNA can be digested simultaneously or consecutively with two enzymes. The banding pattern obtained should allow the unambiguous

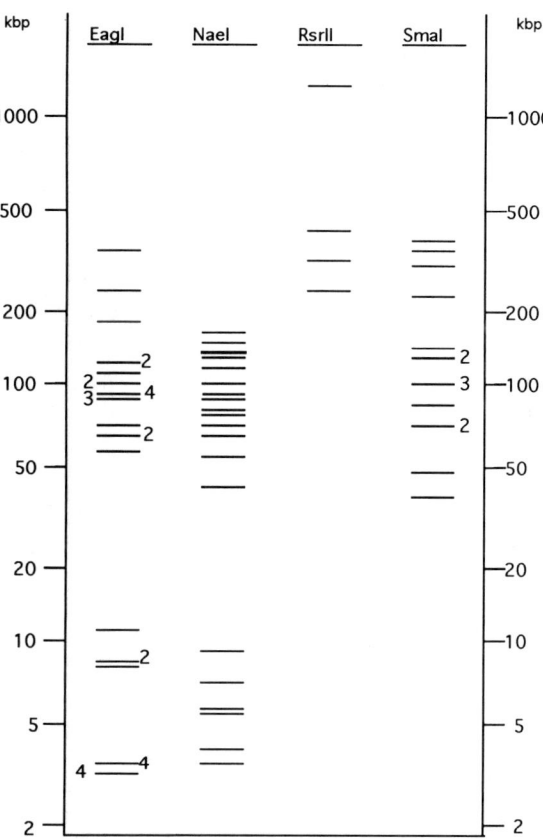

Figure 2. Idealized fractionation of *H. influenzae* DNA digested with *Eag*I, *Nae*I, *Rsr*II, or *Sma*I. Migration of the bands is a function of the logarithm of the size of the fragments. Bands in the *Eag*I and *Sma*I digests are numbered to show the multiplicity of bands where relevant.

sorting of the positions for many of the sites for the two enzymes with respect to each other. If a single-site enzyme is available, such as *Not*I in the example above, then this site can be used as a fixed reference point to orientate fragments from other digests. Problems increase with the number of bands produced and when repeat restriction digestions produce variable banding patterns. These would normally be due to incomplete digestion and can often be confirmed by the presence of variable and faint, abnormally high molecular weight bands towards the top of the relevant gel lane. Incomplete digestion can result from protein remaining in the agarose blocks or from incomplete washing of the reagents used in preparation. Partial digestion can cause mapping anomalies if selected restriction enzyme sites within the genome are more prone to miscutting than others. Deliberate incomplete digestion of DNA can, however, be of use as described below (Section 4.3).

Once basic mapping data have been compiled this should confirm that the chromosome is circular and establish a framework to allow finer mapping with the more frequently cutting enzymes. This must rely on hard work. Some sites can often be mapped from simple double digestion of DNA with a more frequent and a less frequent cutting enzyme but often the restriction patterns become too complex, especially with respect to overlapping bands. The resolution of multiple coincident bands may be assisted by scanning gels or photograph negatives with a densitometer to estimate relative DNA intensity. Another way to proceed is to isolate DNA from individual bands then re-digest these with a second enzyme. This helps to order the sites of the second enzyme within the fragments but care must be taken as repeat digestion often leads to smeary bands and difficult-to-interpret gels.

4.3 Hybridization

One proven way to resolve mapping uncertainties is to isolate DNA from individual bands on gels then use these to make hybridization probes to identify bands with the same sequence in other restriction enzyme digests. In order to do this the pulsed field gel must first be transferred to a solid matrix support. Any procedure describing the Southern transfer of DNA from agarose gels on to membranes is generally applicable to large DNA fragments. *Protocol 5* is our preferred method for blotting CHEF gels.

Protocol 5. Blotting of pulsed field gels

Equipment and reagents

- 0.25 M hydrochloric acid
- Denaturing solution: 0.5 M NaOH, 1.5 M NaCl
- Neutralizing solution: 1.5 M NaCl, 0.5 M Tris–HCl pH 7.4
- 2 × SSC: 0.3 M NaCl, 0.03 M Na$_3$ citrate
- Nylon membrane (e.g. Hybond N, Amersham) or nitrocellulose (e.g. Hybond C, Amersham)
- Short-wavelength UV light box or cross-linking box (Stratagene)

Method

1. Following electrophoresis and band visualization, soak the gel for 20 min in 0.25 M hydrochloric acid with very gentle shaking.[a] This treatment partially depurinates the DNA, causing strand breakage, and allows more efficient transfer of the highest molecular weight DNA.

2. Soak the gel for two 20 min periods (one change) in denaturing solution with gentle shaking.

3. Replace with neutralizing solution and soak the gel for two 20 min periods (one change) with gentle shaking.

---

4. Transfer DNA to a solid support such as nylon membrane or nitrocellulose. For capillary transfer, a standard Southern blot type apparatus can be set up (11) allowing transfer overnight. Vacuum transfer or electroblotting can also be used according to the manufacturers' instructions. Extended transfer times may be required.

5. After transfer rinse the membrane in 2 × SSC, blot dry, then fix the DNA to the filter: UV cross-link for nylon membranes, or bake at 80°C under vacuum for nitrocellulose.

---

[a] Acid hydrolysis can be replaced by UV treatment of ethidium bromide-stained gels.

Carry out hybridization using radiolabelled or non-radiolabelled probes by any suitable method with conditions calculated for the particular DNA. After hybridization, strip the filters and re-probe up to ten times without significant loss of signal.

Following hybridization experiments, calculate the sizes of hybridizing bands for each probe and in this way build up information on the overlapping DNA fragments from different restriction enzyme digests. Also, calculate the number of coincident DNA species in any one gel band. This is a labour-intensive process but does become easier as detail on the restriction map is built up. Compilation of the data may be made easier by use of computer restriction analysis programs.

Another tool which can be used in the restriction analysis of bacterial chromosomes is the directed partial digestion of DNA. Under-digest the DNA (e.g. for 10 h) using restriction enzymes which do not produce too many fragments, fractionate it by PFGE, then hybridize it with a particular DNA fragment or single gene probe. This will identify the target DNA band and most likely several larger bands which are the products of two or more adjacent restriction fragments. The construction of the physical map is aided by the overlapping data from such experiments.

Gaps or ambiguities in the map may arise if small fragments for any particular restriction enzyme (typically less than 20 kb) occur consecutively along the chromosome. These will often be lost off the bottom of pulsed field gels and not be detected. To help overcome this problem, digest high molecular weight DNA prepared by standard methods, fractionate it on non-pulsed field gels, then determine the frequency of lower molecular weight species. Isolate these fragments and use them to make hybridization probes against blots of pulsed field gels to ascertain their positions with respect to other digests (5).

By a combination of these methods, a map has been constructed for the *H. influenzae* strain Eagan chromosome (*Figure 3*) for the enzymes *Not*I, *Nae*I, *Rsr*II, *Sma*I, and *Eag*I.

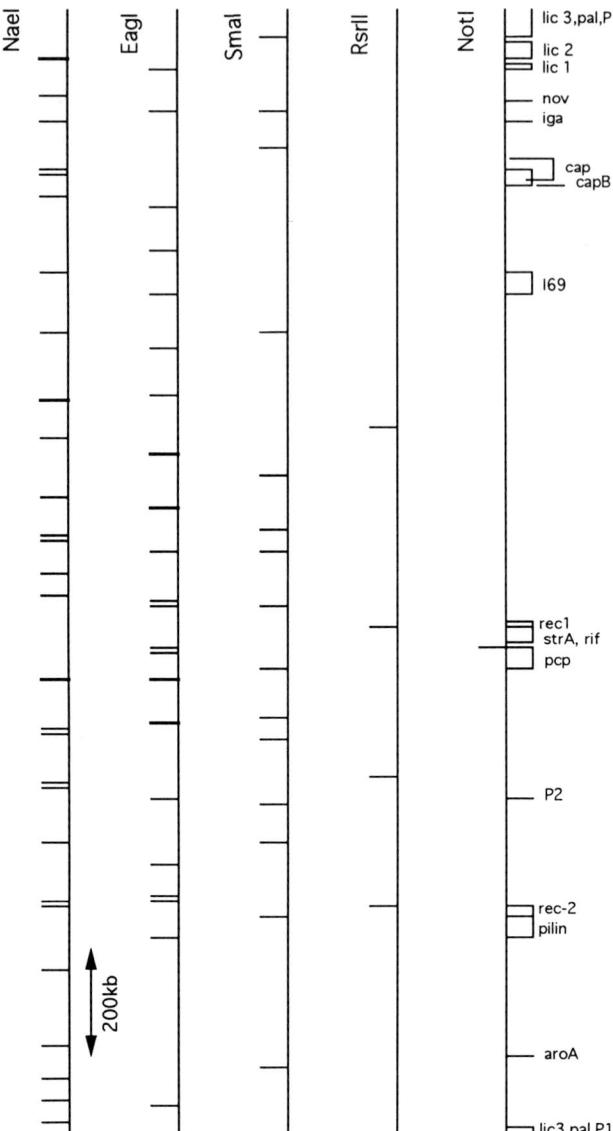

Figure 3. Genomic map of *H. influenzae* type b strain Eagan. The relative positions of restriction sites for the enzymes *Nae*I, *Eag*I, *Sma*I, *Rsr*II, and *Not*I are shown. The positions of the genes given in the text or listed in reference 5 are marked. The total genome length is 2100 kb.

5. Compilation of a genetic map

5.1 Map position of known genes

Once a rudimentary map has been established, use genes which have been cloned and characterized as hybridization probes to map their approximate position and thereby construct a genetic map over the physical restriction map. Single gene probes can also help to confirm some of the finer detail of the restriction map. Small gene size DNA fragments would be expected to hybridize to only one fragment in each of the major genomic digests and thus can accurately confirm the positions of coincident fragments on the map. If the restriction mapping has been completed one would expect to be able to map the position of any given gene with accuracy using single and double enzyme digests of the chromosomal DNA. Check sequenced genes, or cloned DNA with restriction maps, for sites for any of the enzymes used to construct the map. Probes spanning rare-cutting restriction sites are extremely useful as they identify adjacent genomic fragments and promote fragment ordering. For particularly large genomes, prepare a series of linking fragments containing rare recognition sites for the restriction enzymes used to generate the map, as described by Keiser *et al.* (9). These provide spanning probes to link adjacent fragments.

In this manner a genetic map can be drawn to superimpose on the chromosomal restriction map and genes in adjacent positions can be readily identified. Of particular use can be to map the position of the ribosomal RNA (rRNA) gene clusters. They are important repeated sequences in most bacterial chromosomes and from their positions the origin of replication can often be inferred (4).

In *H. influenzae*, rRNA genes also provide a warning in genome mapping experiments. Sequences such as the rRNA operons, which are highly conserved, can have aberrant %G+C contents (4). This can also arise after transfer of DNA segments from other organisms. In *H. influenzae* strain Rd, fine mapping of the number and position of rRNA genes proved difficult because of the relatively high frequency of G/C-rich enzyme sites. If such cases are common it may be necessary to include an atypical restriction enzyme which has no sites within the problem region.

5.2 Pathogenicity genes in *H. influenzae*

As more information becomes available it is becoming apparent that related genes can be clustered on an arc of the chromosome. Our main interest is in the loci involved in the production of the virulence determinants of capsular polysaccharide and lipopolysaccharide in *H. influenzae* strains. Strain Eagan has been used extensively to investigate the role of capsule as a virulence factor. The location of the capsule biosynthetic locus (*cap*B) was determined using a 1.8 kb probe which is known to include some of the biosynthetic genes

(5, 12). The *cap*B locus has been mapped to the left-hand side of the chromosome map (*Figure 3*).

Of particular interest to us are genetic loci (the *lic* loci) which are known to be involved in the reversible phase switching of lipopolysaccharide structural determinants in *H. influenzae* (7). Phase-variable expression of lipopolysaccharide epitopes is likely to be important in optimizing the potential virulence of the organism such that it can express the most appropriate lipopolysaccharide structure for any given environment. Three loci—*lic*1, *lic*2, and *lic*3 are known to be involved with this process and the individual genes show no cross homology to each other and most show little homology to any sequence in the database (7). Interestingly, when the *lic* genes were used as probes each hybridized against distinct chromosomal bands but all mapped to the same small arc of the chromosome. Other distinct pathogenicity determinants have been localized to the same small section of the *H. influenzae* genomic map, namely an outer membrane protein (P1) and a lipoprotein (pal) (5) (*Figure 3*). This supports the notion that pathogenicity determinants can be grouped and closely linked on the chromosome in a way that may be important for their function.

Once such results have been confirmed, the organization of any particular arc of the chromosome can be studied in fine detail using overlapping clones from cosmid or phage clone banks.

6. Maps from closely related strains

Chromosome physical and genetic maps can be used as one of the most complete methods for comparison of strains and to estimate their divergence. The population structure of naturally occurring isolates of capsulate *H. influenzae* shows substantial diversity and comprises many distinct families of strains when characterized for readily determined phenotypes. PFGE can potentially extend the information from the genomic map of a central, well characterized strain to investigate other members of the species. When required, the methods described in *Protocol 1* can be scaled down to allow more rapid processing of multiple strains for comparison by PFGE. 10 ml cultures of bacteria can be treated to give a single agarose block and sufficient DNA for several digestions. Even the most rudimentary maps, perhaps even with a single restriction enzyme, can allow some comparison of gross chromosome structure. If genotypic information is available then gene arrangements between different strains can also be studied.

H. influenzae strains are characterized firstly by their capsular serotype (a–f) and then are split into clusters and families principally by enzyme electrotyping (13). For example, DNA from four *H. influenzae* type b strains from the A1 cluster, including strain Eagan, was digested with the enzyme *Sma*I then analysed by PFGE (5). All gave similar patterns. Three strains from the next most closely related type b group, A2, showed banding patterns which differed from the A1 cluster but still showed a good degree of similarity with each other. Strains from the next family of strains gave unique patterns and in general it was found that banding patterns were radically different between the strains of different serotypes and between random clinical isolates. The *Sma*I patterns of five clinical isolates of different serotypes of *H. influenzae* were distinct from each other and from reference strain Rd at most enzyme sites (4). The size of the *H. influenzae* strain Eagan genome is 2100 kb (5). This is greater than the size determined for another *H. influenzae* strain, Rd. The size of the genome for strain Rd was calculated at 1900 kb (4). Based on serotyping and metabolic enzyme electrotyping (13), strains Rd and Eagan are known to be closely related. However, comparison of the restriction fragment patterns obtained with enzyme *Sma*I shows these strains to be fairly dissimilar (4, 5). Similar disparities in the banding patterns between related strains have been found within many of the other bacterial genera which have been studied. Therefore, caution must be exercised to limit the divergence of strains to be studied if meaningful comparisons are to be made by restriction maps alone. Different restriction patterns for related strains need not mean that their genomes are grossly dissimilar. Restriction site polymorphisms are common and follow from basal mutagenic rates of bacterial DNA. These can be found along with small chromosome deletions, insertions and rearrangements in related genomes. A combined comparison using both restriction and genetic maps would be more useful in these circumstances.

Restriction fragment length polymorphisms have proved useful when comparing genomic maps in *H. influenzae*. Capsule-deficient strains, derived from capsulate strains, have been shown to contain associated deletions of 17 kb of chromosomal DNA which can be detected by a shift in the size of the respective restriction fragments on pulsed field gels (5).

7. Mapping transposon insertion mutants

Transposon mutagenesis is a strategy commonly employed to create random mutations in the genome of a target organism. This is of particular use in identifying unknown genes which code for or influence detectable phenotypes. With *H. influenzae*, the phenotype can be cellular structures such as lipopolysaccharide and capsule where variants can be detected by monoclonal antibody reactivity. This mutagenesis technique can therefore be used to discover novel genes involved with production or expression of pathogenic determinants.

In *Haemophilus*, the transposon Tn916 has proved useful as it has a preference for inserting into A/T base pair rich DNA (14). Random mutagenesis of *H. influenzae* type b strains with Tn916 produced a range of mutants with altered LPS phenotypes, detectable by changes in the interaction of epitopes with specific monoclonal antibodies (N. High, personal communication). The DNA from such mutants can be prepared, digested, run on pulsed field gels, and the position of the transposon determined. Either the transposon can be used as a hybridization probe or larger transposon inserts can be detected as restriction fragment length polymorphisms by band shifts on gels. This allows the investigator to determine how diverse a set of mutations has been obtained. It may be that they have clustered to novel or known arcs of the chromosome or that a set of diverse mutational insertions has been obtained. Fine mapping can reveal whether inserts have occurred in known gene clusters. In an analogous fashion, this type of mapping can be used to map the insertion points of prophage and screen for their presence by band shifts on pulsed field gels.

Acknowledgements

This chapter is dedicated to Dr Peter Butler, research scientist in the Department of Paediatrics 1987–1991, who died on the 18 September 1991. The mapping of strain Eagan was carried out by P. Butler in the laboratory of E. R. Moxon, supported by grants from the Medical Research Council.

References

1. Smith, C. L., Econome, J. G., Schutt, A., Klco, S., and Cantor, C. S. (1987). *Science*, **236**, 1448.
2. Zwahlen, A., Rubin, L. G., and Moxon, E. R. (1986). *Microbiol. Pathogen.*, **1**, 465.
3. Zwahlen, A., Kroll, J. S., Rubin, L. G., and Moxon, E. R. (1989). *Microbiol. Pathogen.*, **7**, 225.
4. Lee, J. J., Smith, H. O., and Redfield, R. J. (1989). *J. Bacteriol.*, **171**, 3016.
5. Butler, P. D. and Moxon, E. R. (1990). *J. Gen. Microbiol.*, **136**, 2333.
6. Anderson, P., Johnson, R. B., and Smith, D. H. (1972). *J. Clin. Invest.*, **51**, 31.
7. Moxon, E. R. and Maskell, D. (1992). In *Molecular biology of bacterial infection* (SGM symposium 49) (ed. C. E. Hornache, C. W. Penn, and C. J. Smythe), p. 75. Cambridge University Press, Cambridge.
8. Herriot, R. M., Meyer, E. M., and Vogt, M. (1970). *J. Bacteriol.*, **101**, 517.
9. Keiser, H. M., Keiser, T., and Hopwood, D. (1992). *J. Bacteriol.*, **174**, 5496.
10. Chu, G., Vollrath, D., and Vogt, M. (1970). *Science*, **234**, 1582.
11. Sambrook, J., Fritsch, E. F., and Maniatis, T. (1989). In *Molecular cloning: a laboratory manual* (2nd edn). Cold Spring Harbor Press, Cold Spring Harbor, NY.
12. Ely, S., Tippett, J., Kroll, J. S., and Moxon, E. R. (1986). *J. Bacteriol.*, **167**, 44.
13. Musser, J. M., Kroll, J. S., Granoff, D. M., Moxon, E. R., Brodeur, B. R., Campos, J., *et al.* (1990). *Rev. Infect. Diseases*, **12**, 75.
14. Kauc, L. and Goodgal, S. H. (1989). *J. Bacteriol.*, **171**, 6625.

9

Analysis of the genomes of protozoan parasites using PFGE

DAVID J. KEMP and ROBERTO CAPPAI

1. Introduction

Pulsed field gel electrophoresis (PFGE) has played a major role in developing our current knowledge of the genomes of a number of protozoan parasites as the genomes are small and the chromosomes usually fall within a usable size range. The studies on malaria and trypanosomatids described here demonstrate the applicability of the techniques to other protozoa.

2. Malaria

Most work has concentrated on the important human pathogen *Plasmodium falciparum* and the discussion here will be limited to this organism.

2.1. *P. falciparum* chromosomes

Classical genetic analysis of the malarial parasite *P. falciparum* is an extremely difficult process because of the complexity of the life cycle between humans and mosquito. As the chromosomes do not form recognizable mitotic figures it was not even known how many chromosomes there were until the advent of PFGE. It has now become clear that there are 14 chromosomes and markers exist for all of these (1–5). Several of the chromosomes have been mapped using rare-cutting restriction endonucleases. As the genome of *P. falciparum* is remarkably rich in A+T (82%) a considerable number of enzymes are useful. A number of mechanisms that result in the chromosomes varying in size quite considerably from isolate to isolate have been clarified. These mechanisms include subtelomeric deletions that occur by a breakage–healing mechanism and amplifications of some regions. Libraries of yeast artificial chromosomes (YACs) have been constructed (6, 7) and an international consortium involving laboratories in the UK, Australia, and USA is currently generating a YAC map of the entire 25 Mb genome. Chromosomes 2 and 4 have already been completely spanned by YACs. Several groups are currently using positional information as an approach to understand various aspects of the biology of *P. falciparum*, an approach that was not conceivable ten years ago, and the YAC map will be of great value for this.

Most commonly, *P. falciparum* chromosomes are prepared from parasites cultured on red blood cell preparations that contain few white cells. *Protocol 1* describes the preparation of *P. falciparum* chromosomes from cell cultures. For preparation of chromosomes from fresh blood samples taken directly from patients rather than cells in culture, additional steps are necessary to remove white blood cells. These procedures are described in *Protocol 2*.

Protocol 1. Preparation of *P. falciparum* chromosomes from cultured parasites[a]

Equipment and reagents

- *P. falciparum* culture, approximately 5–10% parasitaemia. The reference isolate for chromosomes is clone 3D7 which can be obtained from Dr David Walliker, Department of Genetics, University of Edinburgh.
- RPMI–Hepes containing 0.15% (w/v) saponin
- Phosphate-buffered saline (PBS)
- 2% (w/v) LMT (low melting temperature agarose) in PBS
- Storage buffer: 50 mM EDTA, 0.01 M Tris–HCl pH 8.0

- Bench centrifuge
- Mould for casting agarose plugs (as described in Chapter 2). If the DNA is to be cut subsequently with restriction endonucleases it is important that the agarose is free of inhibitory material. Agarose suitable for this is available from BioRad, FMC, and Pharmacia. Melt it in a microwave oven and cool to 42–45°C immediately prior to use.
- Lysis buffer: 0.5 M EDTA, 0.01 M Tris–HCl pH 8.0, 1% (w/v) sarkosyl containing 2 mg/ml Proteinase K

Method

1. Resuspend the culture. Pellet the red blood cells (5 min, 1400 *g*, bench centrifuge) and note the packed cell volume (PCV).

2. Resuspend the pellet in 1.5 volumes (PCV) of RPMI–Hepes containing 0.15% (w/v) saponin and leave on ice for about 5 min.

3. Pellet the parasites (5 min, 12 000 *g*).

4. Discard *all* the supernatant (using a transfer pipette). Note the packed parasite volume (PPV).

5. Resuspend the parasite pellet in 1.5 volumes (PPV) of PBS and mix.

6. Add an equal volume (i.e. 2.5 × PPV) of 2% (w/v) LMT agarose in PBS (molten and cooled to 42–45°C).

7. Pipette the mixture into the mould (pre-chilled on ice) and allow the blocks to set (on ice).

8. Push the blocks out of the mould into lysis buffer containing 2 mg/ml Proteinase K (allow 1 ml/5 blocks) and incubate at 42°C for 2 days.

9. Store the blocks at 4°C in storage buffer.

[a] From ref. 8.

Parasites in blood samples taken directly from patients are almost entirely at the ring stage of development, containing 1 genome per parasite. The parasites can be cultured for 36–48 h until they have completed replication of DNA but the schizonts have not ruptured. This provides the maximum amount of DNA but as each parasite is still within the same red cell, there has been no chance for selection *in vitro*.

Protocol 2. Preparation of chromosomes from *P. falciparum* taken from natural parasite populations[a]

Equipment and reagents

- Blood sample (heparinized)
- RPMI–Hepes
- 5% (w/v) aqueous sorbitol
- 10% human group O serum in RPMI–Hepes
- Gas mixture (1% O_2, 5% CO_2)
- Bench centrifuge

Method

1. Pellet the parasitized red blood cells (5 min, 1400 *g*).

2. Remove the buffy coat.

3. Resuspend the sample in 5% (w/v) aqueous sorbitol for 10 min at room temperature.

4. Pellet the sample again, remove the buffy coat and resuspend in RPMI–Hepes.

5. Pellet the sample again, remove the buffy coat, and resuspend at 6% haematocrit in 10% group O serum.

6. Incubate at 37°C in 1% O_2, 5% CO_2 for 36–48 h until the majority of the cells are schizonts or trophozoites.

7. Pellet the sample again and resuspend in RPMI–Hepes.

8. Follow from step 1 of *Protocol 1*.

[a] From ref. 9.

2.1.1 Assignment of cloned DNA to chromosomes by PFGE

The methods here are the same as for other PFGE experiments. The parameters for separation in a particular apparatus depend on the size range of the chromosomes. For a CHEF apparatus as originally described (10) the following conditions are suitable for separation of all *P. falciparum* chromosomes:

- 270 sec pulse, 44 h, 120 V, followed by
- 1800 sec pulse, 24 h, 70 V.

Chromosomes 13 and 14 can be separated much better (11) by

- 999 sec pulse, 144 h or longer, 70 V.

In a CHEF-DRIII machine (BioRad) use the following conditions to separate all but the largest chromosomes:

- 300 sec pulse, 72 h, 4 V/cm, field angle 106°.

This field angle allows higher DNA loads than at 120° and is therefore ideal for preparative work (12).

We have found to our surprise that some batches of agarose that give good results for separation of yeast chromosomes give badly smeared separations of *P. falciparum* chromosomes. Sections cut from the same chromosome blocks can behave quite differently on different batches of agarose. We currently use Seakem (FMC) agarose but each batch must be tested.

The isolate chosen as the standard for *P. falciparum* chromosomes is 3D7, as it is a clone that has been shown to be capable of traversing the entire life cycle and most of its chromosomes are readily separated (1, 4). A combination of clones D10, E12, and 3D7 give unique patterns for each chromosome (1). Even with these, use chromosome-specific markers as controls. With other isolates it is very unlikely that all chromosomes will run in the same numerical order because of size variations. A suitable procedure is described in *Protocol 3*.

Protocol 3. Blotting *P. falciparum* chromosomes for assignment of genes[a]

Equipment and reagents

- PFGE apparatus
- Polaroid camera and film
- Hybond-N nylon filters (Amersham)
- Ethidium bromide: 0.5 μg/ml in 0.5 × TBE
- Denaturing solution: 0.5 M NaOH, 1.5 M NaCl
- Neutralizing solution: 0.5 M Tris–HCl pH 7.4, 1.5 M NaCl

- Hybridization mix: 6 × SSC, 5 × Denhardt's solution, 0.5% (w/v) SDS, and 500 μg/ml herring sperm DNA as carrier. 1 × Denhardt's solution is 0.02% (w/v) BSA, 0.02% (w/v) Ficoll, and 0.02% (w/v) polyvinylpyrolidone. 1 × SSC is 0.15 M NaCl, 15 mM Na citrate, pH 7.0.
- Long-wave UV light box (or transilluminator)
- Short-wave UV light box (or transilluminator)

Method

P. falciparum chromosomes are separated by PFGE under the conditions described above.

1. Stain the gel with ethidium bromide (approximately 1 h, room temperature).

2. Photograph the gel. If there is enough DNA, photograph the gel immediately on the long-wave UV box. Alternatively, destain the gel for about 2 h and then photograph it on the short-wave UV box.

3. Expose the gel to short-wave UV light to fragment the DNA (3 min).

4. Rock the gel in denaturing solution (40 min).

5. Rock the gel in neutralizing solution (2 × 30 min).

6. Transfer the DNA to the nylon filter by standard semi-dry blotting procedure.

7. Fix the DNA to the nylon filter by exposing to the short-wave UV light for 3 min, or according to the manufacturers' instructions.

8. Pretreat the filter in hybridization mix for at least 1 h.

9. Add the relevant probe, incubate overnight, and wash by standard methods.

---

[a] The conditions described here are those used for most of the work from our laboratories, e.g. ref. 13. Many other variations are equally effective.

2.1.2 Restriction mapping of *P. falciparum* chromosomes

Because of the extreme A+T content of *P. falciparum* DNA, many restriction endonucleases that recognize six base sites comprised entirely of G+C are useful for mapping chromosomes. Useful enzymes include *Apa*I, *Sma*I, *Bgl*II, *Bss*HII, *Sac*II, *Eag*I, *Not*I, and *Sgr*AI. This has allowed the generation of maps of several entire chromosomes. Restriction digests can be carried out on unfractionated chromosomes embedded in agarose plugs as described in *Protocol 4* if the probe to be used is a unique sequence. If the probe is a multicopy sequence that is located on several or all chromosomes then it is necessary to first fractionate the chromosomes by PFGE and cut out the relevant chromosome as described in *Protocol 5*.

Protocol 4. Restriction mapping of total *P. falciparum* chromosomes [a]

Equipment and reagents

- Restriction enzymes
- *P. falciparum* chromosomes in agarose plugs (see *Protocols 1* and *2*).

- Restriction enzyme buffers. (The buffers recommended by the manufacturers but containing 0.01% (v/v) Triton X-100, 10 μg/ml acetylated BSA, and 0.1 mM DTT are used.)

Method

1. Equilibrate a small agarose block containing chromosomal DNA molecules representing the entire *P. falciparum* genome in 500 μl of the appropriate restriction enzyme buffer for 30 min at room temperature.

2. Replace this buffer with 200 μl of restriction enzyme buffer containing

---

Protocol 4. *Continued*

a 2–5-fold excess of enzyme (usually 20 units) and carry out digestion for 1–3 h at the appropriate temperature.

3. Fractionate the restriction fragments by PFGE:
   - 90 sec pulse, 16 h, 160 V.

4. Blot the gel and hybridize as described in *Protocol 3*.

[a] From ref. 13.

Protocol 5. Restriction mapping of isolated *P. falciparum* chromosomes[a]

Equipment and reagents

- Long wavelength (360 nm) UV light
- Scalpel blades
- 50 mM EDTA, 10 mM Tris–HCl pH 8.0
- Butanol saturated with 1 M NaCl, 10 mM Tris pH 8.0, 1 mM EDTA at room temperature

Method

1. Stain a preparative gel of intact chromosomes with ethidium bromide and visualize under long wavelength UV light. Do not use a short wavelength UV light since it fragments the chromosomes.

2. Excise the chromosome bands from the gel in agarose blocks that are as small as possible with a scalpel blade.

3. Remove the ethidium bromide by butanol extraction (4 × 30 min wash).

4. Store the purified chromosomes at 4°C in 50 mM EDTA, 10 mM Tris–HCl pH 8.0.

5. Carry out the restriction enzyme digestion as described in *Protocol 4*.

[a] From ref. 13.

2.1.3 Two-dimensional PFGE for detecting subtelomeric deletions of *P. falciparum* chromosomes

In cultured isolates of *P. falciparum*, deletion of DNA from subtelomeric regions is very common. This is of considerable interest because the genes for many specialized parasite functions are located in these regions. The correlation of loss of such functions with specific deletions therefore provides an approach for using positional information to clone the relevant genes.

The ends of undeleted *P. falciparum* chromosomes have a common structure. As for all eukaryotes they terminate in telomeric repeats. Internal to this is a region about 10–20 kb long containing at least one *Apa*I site, and proximal to this is a large block of non-transcribed repetitive DNA designated rep 20.

This common structure means that all telomeric *Apa*I fragments originating from undeleted chromosomes are small (~10–20 kb). As the deletions occur by chromosome breakage with subsequent healing (i.e. addition of new telomeric repeats), telomeric *Apa*I fragments originating from deleted regions will be about the size of other chromosomal *Apa*I fragments, averaging several hundred kb. This has allowed the development of a two-dimensional PFGE strategy for detection of subtelomeric deletions occurring anywhere in the genome on the one gel.

Fractionate the chromosomes first by PFGE using appropriately long pulses. Cut out a strip about 2 mm square from the origin through the length of the gel and treat it with *Apa*I. Load the strip across the origin of a second gel and separate the *Apa*I fragments by PFGE with considerably shorter pulses, blot, and hybridize with a telomeric probe.

Telomeric fragments from undeleted chromosomes occur as spots across a fast-moving 'front' while the much larger fragments from deleted chromosomes are clearly evident behind this (*Figure 1*). Hybridization with the rep20 probe helps confirm that deletions have occurred because the spots corresponding to these regions disappear. The procedure is described in *Protocol 6*.

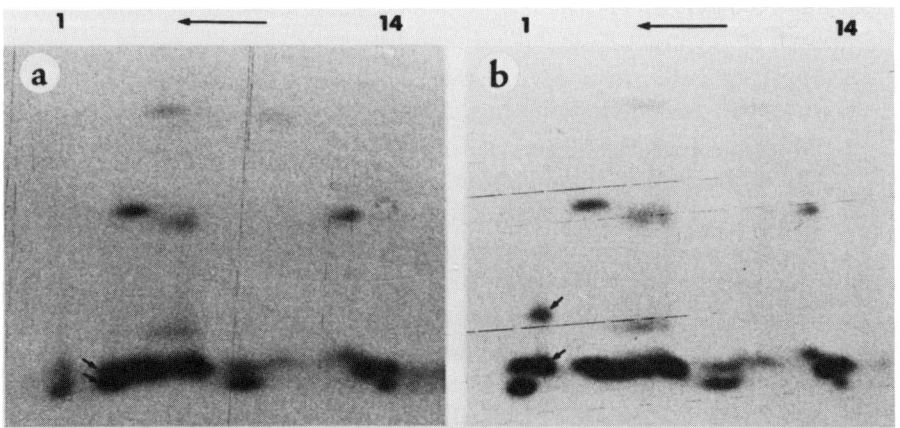

Figure 1. Two-dimensional PFGE of variant ItG2 F6 clones. (a) Knob-positive cytoadherent clone 3. The arrows show two small telomeric *Apa*I fragments of chromosome 2. (b) Knobless non-cytoadherent clone 6. The arrows show that one of the two telomeric fragments is much larger while the other is the same size as for clone 3. Because of the deletion on chromosome 2, both are closer to the telomeres of chromosome 1 than for clone 3 (23) (kindly provided by B. A. Biggs).

Protocol 6. Two-dimensional PFGE for detecting subtelomeric deletions of *P. falciparum* chromosomes[a]

Equipment and reagents

- Scalpel blade
- Plastic ruler
- Parafilm strip, 3 cm wide, 2 cm longer than gel strip, creased down the middle
- Sticking tape
- Empty pipette tip box
- PFGE apparatus

Method

1. Fractionate the chromosomes by PFGE, e.g. using
   - 270 sec pulses, 44 h, 120 V followed by
   - 1800 sec pulses, 24 h, 70 V.

2. Stain the gel with ethidium bromide (0.5 μg/ml).

3. Cut a strip approximately 2 mm wide through the length of each track required, using a long-wave UV box to illuminate the gel and a scalpel and plastic ruler to cut it.

4. Lay the strips on their sides on the long-wave UV box. Cut off the regions that were originally the top and bottom surfaces of the gel, using the scalpel and plastic ruler, leaving a strip approximately 2 mm square containing the DNA. Cut off the bottom of this at a 45° angle so that it can be distinguished from the top (cut off square at the origin).

5. Place the strip on a piece of parafilm 3 cm wide and about 2 cm longer than the gel, which has already been creased down the middle (long axis) so it can easily be folded.

6. Fold the parafilm in halves over the gel. Seal each end by pressing down with the back of a pair of scissors (keep them well clear of the gel).

7. With a piece of sticking tape at each end, stick the folded piece of parafilm to the vertical side of a plastic box (e.g. a pipette tip box). It now forms a narrow trough, open at the top with the gel strip at the bottom.

8. Follow steps 3–5 of *Protocol 5*. Approximately 1–2 ml of solution is necessary to submerge the gel strip depending on its exact size. To incubate, place the box with strip attached in an air incubator.

9. Cast a gel with a sufficiently wide well to take the strip (using the back of the comb is one way). Load markers at each side. Seal in the gel strips with molten LMT as usual.

10. Separate the fragments by PFGE:
    - 20 sec pulses, 19 h, 160 V.

11. Stain, photograph, blot the gel, and hybridize the DNA with a telomere probe as in *Protocol 3*.

3. *Leishmania*

Leishmania species are trypanosomatids which exhibit several unusual molecular phenomena including trans-splicing and RNA editing. The PFGE technique has allowed a number of characteristics of the *Leishmania* genome to be defined. It has shown that the number and size of the *Leishmania* chromosomes vary between species, isolates, and clones. A feature of PFGE separated *Leishmania* chromosomes is that they exhibit non-stoichiometric ethidium bromide-staining between chromosomes, due to a variable number of similar or different chromosomes of the same size.

The following sections will describe the application of the PFGE technique in *Leishmania* for chromosome assignment of cloned DNA and for epidemiology and diagnostic studies. *Leishmania* has been a useful model system to study mechanisms of drug resistance. The use of PFGE for the detection and correlation of DNA amplicons with drug resistance is outlined.

3.1 Preparation of *Leishmania* chromosomes

The majority of PFGE studies on *Leishmania* have involved analysis of chromosomes from promastigotes due to their ease of growth *in vitro* in a variety of commercially available tissue culture media (i.e. Schneider's Drosophila medium) supplemented with fetal calf serum. A comparison of the karyotype between the amastigote and promastigote stages has shown no differences, thus supporting the validity of using promastigote chromosome blocks. The preparation of *Leishmania* chromosome blocks is similar to that used for other organisms and is shown in *Protocol 7*.

Protocol 7. Preparation of *Leishmania* chromosome blocks [a]

Equipment and reagents

- Bench centrifuge
- PBS (*Protocol 1*)

Method

1. Wash promastigotes three times in PBS.

2. Resuspend promastigotes in PBS at 2.5×10^8 cells/ml. [b]

3. Follow steps 6 to 9 in *Protocol 1*.

[a] From ref. 14.
[b] The parasite concentration in the agarose blocks can affect the resolution of the chromosomes and should be optimized for the PFGE apparatus to be used.

3.2 Assignment of cloned DNA and genes

The gel running conditions on a BioRad CHEF-II mapper required to separate the small to medium sized chromosomes are

- 60 sec pulse, 24 h, 6 V/cm, followed by
- 80 sec pulse, 24 h, 6 V/cm.

Run conditions to separate the larger chromosomes are

- 3600 sec pulse, 72 h, 3.5 V/cm.

Process the gels for ethidium bromide staining and Southern blotting and probe with the gene of interest as described in *Protocol 3*.

The procedure for analysing *Leishmania* chromosome structure is the same as that described for *Plasmodium* chromosomes (see Section 2.1.4). Separate the chromosomes by PFGE, excise from the gel, digest with various restriction endonucleases, and analyse the fragments by PFGE and Southern blotting. However, as the genome is G+C rich, the choice of enzymes is different. Such studies suggest that as for malaria chromosomes, the subtelomeric region is responsible for the majority of the polymorphisms observed.

3.3 Application to epidemiology and diagnosis

PFGE provides a powerful technique to compare the relatedness between a parent and its clones (i.e. clonality) based on their karyotype pattern. This procedure can also be used to classify *Leishmania* species from the field where classification involves comparing the karyotype of the field isolate against a panel of known *Leishmania* species. Separated chromosomes are analysed by ethidium bromide staining pattern and/or blotted and probed with appropriate gene markers (such as α, β-tubulin). However, given the high degree of chromosomal polymorphisms observed among members of the same species and subspecies, caution must be exercised in trying to correlate chromosome band sizes to a particular species (14). PFGE has been used to determine if a recrudescence of *L. braziliensis* disease in humans is due to reactivation of persistent infections or due to exogenous reinfection. It has also been used to show that *L. major* parasites which persist in mice one year after recovery are identical to the parent strain (15).

3.4 Analysis of amplicons

Leishmania has been used as a model system to study mechanisms of drug resistance. Usually, drug resistance is conferred by gene amplification where the amplified DNA is present as large (30–80 kb) extrachromosomal circles. Amplified DNA is usually detected by digesting total *Leishmania* DNA with restriction enzymes and analysing the products on an ethidium bromide-stained agarose gel. The amplified DNA is detected as prominent bands not present in the non-resistant (wild-type) parent.

A number of criteria are used to establish that the amplified DNA is circular and not present on linear DNA. The majority of these procedures use PFGE to analyse the DNA. These criteria are: pulse-time dependence of migration in PFGE, whether the DNA can be extracted by alkaline lysis, susceptibility to λ-exonuclease digestion, change of mobility after γ irradiation, and hybridization to a telomere probe. The study of α-difluoromethylornithine-resistant *L. donovani* illustrates most of these techniques (16). The methods are outlined below.

3.4.1 Pulse-time-independent migration of circular DNA

The migration of circular molecules by PFGE has been shown to be pulse-time independent (17). That is, for a given voltage/cm and run time, a circular molecule will travel approximately the same distance with different pulse times. The distance travelled by linear DNA is dependent on pulse time. Therefore, *Leishmania* DNA is separated by PFGE for the same run time and voltage but for different pulse times and the migration of the amplified DNA is measured either by ethidium bromide staining and/or Southern blotting with a specific probe.

3.4.2 Mobility and sizing of circular DNA molecules by γ irradiation

The introduction a single random double-stranded break with limited γ irradiation converts supercoiled DNA into linear DNA (18). Therefore, if γ irradiation results in a dose-dependent conversion of a DNA molecule into a new discrete band this indicates the DNA is circular. The DNA is irradiated in agarose with a ^{60}Co source for various doses as described in *Protocol 8*. A dose of 20 krad of γ radiation introduces one double-stranded break per 500 kb using the buffer below. This approach has been used to accurately size the H region of multiple drug resistant *L. major* (19).

Protocol 8. Gamma irradiation

Equipment and reagents

- ICN GR9 irradiator using a ^{60}Co source
- All washes and irradiation are done at room temperature in 1.5 ml microcentrifuge tubes
- *Leishmania* chromosomes in agarose plugs (see *Protocol 7*)
- 0.2 M Tris–HCl pH 7.4/0.1 M EDTA

Method

1. Wash *Leishmania* chromosomes in agarose plugs extensively in 0.2 M Tris–HCl pH 7.4/0.1 M EDTA.

Protocol 8. *Continued*

2. Expose the plugs to increasing doses of γ irradiation (typically 0, 0.6, 2.0, 6.0 and 20.0 krad).

3. Separate the DNA by PFGE and analyse by Southern blotting as in *Protocol 3*.

3.4.3 Sensitivity of DNA to λ exonuclease digestion

The sensitivity of linear DNA to λ exonuclease digestion is a useful criterion to determine if a DNA molecule is linear or circular. The substrate for λ exonuclease is double-stranded DNA with a 5′-terminal phosphate, while nicked circular DNA will not be digested. Therefore, chromosomes (linear, double-stranded DNA) will be digested by λ exonuclease while circular DNA is unaffected. Following λ exonuclease digestion as described in *Protocol 9* the DNA is separated by PFGE and analysed by ethidium bromide staining and/or Southern blotting.

Protocol 9. λ exonuclease digestion for distinguishing circular DNA from linear DNA

Equipment and reagents

- λ exonuclease buffer: 67 mM glycine–KOH (pH 9.4), 2.5 mM MgCl$_2$, 50 μg/ml BSA
- All washes and incubations are done at room temperature in 1.5 ml microcentrifuge tubes
- *Leishmania* chromosomes in agarose plugs (see *Protocol 7*)
- EDTA (0.5 M, pH 8.0)
- PFGE apparatus

Method

1. Wash chromosome agarose plugs twice for 2 h in λ exonuclease buffer minus magnesium, BSA, and enzyme.

2. Wash once in complete λ exonuclease buffer minus enzyme.

3. Add complete λ exonuclease buffer plus λ exonuclease (200 units) in a volume of 400 μl.

4. Incubate at 37°C for various times up to 2 h.

5. Stop reactions by adding EDTA to 50 mM.

6. Analyse by PFGE and Southern blotting as in *Protocol 3*.

---

4. *Trypanosomes*

The PFGE technique has been used in the study of *Trypanosome* karyotypes, the mapping of genes, and the elucidation of chromosomal rearrangements, especially involving the variant surface glycoprotein (VSG). PFGE-separated

Trypanosome chromosomes also exhibit non-stoichiometric ethidium bromide staining between chromosomes.

4.1 Preparation of *Trypanosome* chromosomes and gene assignment

The protocol for preparing *Trypanosome* chromosome blocks is the same as for *Leishmania* (see *Protocol 7*). Studies have shown there is no difference in the chromosome pattern between different life-cycle stages.

A number of studies have reported the chromosomal assignment of cloned *Trypanosoma* genes. Van der Ploeg *et al.* (20) have reported conditions to separate *T. brucei* chromosomes into 20 bands using a Pharmacia LKB Pulsaphor apparatus. The conditions used for the larger chromosomes (up to 5.7 Mb) were

- 4500 sec pulse, 99 V, for 269 h.

To separate chromosomes between 0.3 to over 1.6 Mb use the following conditions:

- 90 sec pulse, 140 V, 24 h followed by
- 200 sec pulse, 112 V, 24 h followed by
- 350 sec pulse, 84 V, 24 h followed by
- 500 sec pulse, 84 V, for 24 h.

4.2 Detection of chromosomal rearrangements

The main class of chromosomal rearrangements that have been studied in *Trypanosoma* have involved the VSG genes. The mechanism(s) controlling VSG expression would have remained difficult to solve without the advent of PFGE. It is known that an expressed VSG is located at the end of the chromosome, and these VSG gene expression sites are located only on a few specific chromosomes.

PFGE has been used to study the movement of a VSG gene from its non-expressed site to a VSG gene expression site on another chromosome (21). Therefore, chromosome blocks are prepared from a variant line before and after it switches to another antigenic variant form (Section 4.1). PFGE Southern blots are then probed with the VSG genes from the 'before' (VSG$_b$) and 'after' (VSG$_a$) variant forms (*Protocol 3*). Such genes are isolated from cDNA libraries made from each variant line. A typical analysis is presented by Shea *et al.* (21).

4.3 Molecular speciation

The PFGE technique has been used for the typing of *Trypanosoma*. It has been shown that *T. congolense* clones of a specific antigenic repertoire (serodeme) have a similar chromosome profile in terms of number of bands and size, while those with different antigenic profiles differ significantly with regard to chromosome size and number (22).

Most of the studies which have compared the karyotypic profiles both within and between species have concluded that there is considerable variation in the size and number of chromosomes as assayed by probing with specific genes. This variation would indicate that typing species on the basis of their chromosome banding profile is difficult.

5. Analysis of transfected kinetoplastids by PFGE

The advent of DNA transfection in kinetoplastids (23) has benefited greatly from the use of PFGE. Depending on the DNA construct, the transfected DNA is either targeted into the genome (usually by homologous recombination) or is maintained extrachromosomally on episomes. These two possibilities are distinguished as follows: embed the transfected parasites into low melting temperature agarose plugs (*Protocol 1*). Separate the DNA by PFGE and probe with a transfection-construct specific probe (usually the drug selection gene; *Protocol 3*). If the transfection construct resides as episomal DNA it will exhibit pulse-time-independent mobility due to its circular structure (Section 3.4.1). Integrated DNA will hybridize to the chromosome which contains the targeted gene. Furthermore, the targeted DNA will exhibit pulse-time-dependent mobility as it resides on a linear DNA molecule.

Acknowledgements

This work was supported by the Howard Hughes Medical Institute, the National Health and Medical Research Council (Australia), and the Wellcome Trust. We thank Colin Sutherland, Peter Bourke, Ivan Bastian, and Deby Holt for helpful suggestions and Gabrielle Falls for help in preparing the manuscript.

References

1. Kemp, D. J., Thompson, J. K., Walliker, D., and Corcoran, L. M. (1987). *Proc. Natl. Acad. Sci. USA*, **84**, 7672.
2. Van der Ploeg, L. H. T., Smits, M., Ponnudurai, T., Vermeulen, A., Meuwissen, J. H. E. T., and Langsley, G. (1985). *Science*, **229**, 658.
3. Wellems, T. E., Walliker, D., Smith, C. L., do Rosario, V. E., Maloy, W. L., Howard, R. J., *et al.* (1987). *Cell*, **49**, 633.
4. Walliker, D., Quakyi, I. A., Wellems, T. E., McCutchan, T. F., Szartman, A., London, W. I., *et al.* (1987). *Science*, **236**, 1661.
5. Triglia, T., Wellems, T. E., and Kemp, D. J. (1992). *Parasitol. Today*, **8**, 225.
6. Triglia, T. and Kemp, D. J. (1991). *Mol. Biochem. Parasitol.*, **44**, 207.

7. DeBruin, D., Landser, M., and Ravetch, J. V. (1992). *Genomics*, **14**, 332.
8. Kemp, D. J., Corcoran, L. M., Coppel, R. L., Stahl, H. D., Bianco, A. E., Brown, G. V., and Anders, R. F. (1985). *Nature*, **315**, 347.
9. Corcoran, L. M., Forsyth, K. P., Bianco, A. E., Brown, G. V., and Kemp, D. J. (1986). *Cell*, **44**, 87.
10. Chu, G., Vollrath, D., and Davis, R. W. (1986). *Science*, **234**, 1582.
11. Karcz, S., Hermann, V. R., and Cowman, A. F. (1993). *Mol. Biochem. Parasitol.*, **58**, 333.
12. Bourke, P. F., Holt, D. C., and Kemp, D. J. (1994). *Trends Genet.*, **10**, 115.
13. Corcoran, L. M., Thompson, J. K., Walliker, D., and Kemp, D. J. (1988). *Cell* **53**, 807.
14. Lighthall, G. K. and Giannini, S. H. (1992). *Parasitol. Today*, **8**, 192.
15. Aebischer, T., Moody, S. F., and Handman, E. (1993). *Infect. Immunol.*, **61**, 220.
16. Hanson, S., Beverley, S. M., Wagner, W., and Ullman, B. (1992). *Mol. Cell. Biol.*, **12**, 5499.
17. Beverley, S. M. (1988). *Nucleic Acids Res.*, **16**, 925.
18. Beverley, S. M. (1989). *Anal. Biochem.*, **177**, 110.
19. Ellenberger, T. E. and Beverley, S. M. (1989). *J. Biol. Chem.*, **264**, 15094.
20. Van der Ploeg, L. H. T., Smith, C. L., Polvere, R. I., and Gottesdiener, K. M. (1989). *Nucleic Acids Res.*,, **17**, 3217.
21. Shea, C., Glass, D. J., Parangi, S., and Van der Ploeg, L. H. T. (1986). *J. Biol. Chem.*, **261**, 6056.
22. Masake, R. A., Nyambati, V. M., Nantulya, V. M., Majiwa, P. A. O., Moloo, S. K., and Musoke, A. J. (1988). *Mol. Biochem. Parasitol.*, **30**, 105.
23. Coburn, C. M., Otteman, K. M., McNeely, T., Turco, S. J., and Beverley, S. M. (1991). *Mol. Biochem. Parasitol.*, **46**, 169.
24. Biggs, B. A., Kemp, D. J., and Brown, G. V. (1989). *Proc. Natl. Acad. Sci. USA*, **86**, 2428.

A1

Addresses of suppliers

American Type Culture Collection, 12301 Parklawn Drive, Rockville Maryland 20852, USA.

Amersham Internationals plc, Amersham Place, Little Chalfont, Buckinghamshire HP7 9NA, UK.

Amicon, Inc., Beverly, MA 01915, USA.

BDH Ltd, Broom Road, Poole, Dorset BH12 4NN, UK.

Beckman Instruments Inc., Spinco Division, 1050 Page Mill Road, Palo Alto, CA 94304, USA.

Becton Dickinson and Company, Becton Dickinson Labware, 2 Bridgewater Lane, Lincoln Park NJ 07035, USA.

Bethesda Research Laboratories, Life Technologies Inc., PO Box 6009, 8717 Grovemont Circle, Gaithersburg, MD 20877, USA; Life Technologies Ltd, PO Box 35, Trident House, Renfrew Road, Paisley PA3 4EF, UK.

Bibby Sterilin Ltd, Stone, Staffs, UK.

BIO 101 Inc., PO Box 2284, La Jolla, CA 92038, USA.

BioRad Laboratories Inc., 2000 Alfred Nobel Drive, Hercules, CA 94547, USA.

Biometra biomedizinische Analytik GmbH, D-3400, Göttingen, Germany; Biometra Ltd, PO Box 167, Maidstone, Kent ME14 2AT, UK.

Boehringer Mannheim GmbH, Biochemica, Sandhofer Strasse 116, 68298 Mannheim, Germany.

Calbiochem Novabiochem, 3 Heathcoat Building, Highsfield Science Park, University Boulevard, Nottingham NG7 2QJ, UK.

Clontech Laboratories Inc., 4030 Fabian Way, Palo Alto, CA 94303–4607, USA.

Difco Laboratories, PO Box 331058, Detroit, MI 48232, USA; PO Box 14B, Central Avenue, East Molesey, Surrey KT8 08E, UK.

DuPont Company, Biotechnology Systems, BRML, G-50636, Wilmington, DE 19898, USA.

Epicenter Technologies, 1202 Ann Street, Madison, WI 53713, USA.

Falcon, *see* **Becton Dickinson and Company**.

FMC Marine Colloids, Bioproducts Department, 5 Maple Street, Rockland, ME 04841, USA.

Hoefer Scientific Instruments, PO Box 77387, 654 Minnesota Street, San Francisco, CA 94107, USA.

Hybaid Limited, 111–113 Waldegrave Road, Teddington, Middlesex TW11 8LL, UK.

ICN Biomedicals, 3300 Hyland Ave, Costa Mesa, CA 92626, USA.

Invitrogen Corporation, 3985 B Sorrento Valley Boulevard, San Diego CA 92121, USA; Invitrogen BV, De Schelp 26, 9351 NV Leek, The Netherlands.

LEEC Ltd, Private Road No. 7, Colwick Estates, Nottingham NG5 2AJ, UK.

Marabuwerke GmbH and Co., Tamm D-7146, Germany.

New Brunswick Scientific, Edison House, 163 Dixons Hill Road, North Mimms, Hatfield AL9 7JE, UK.

New England Biolabs Inc., 32 Tozer Road, Beverly MA 01915, USA; CP Labs, PO Box 22, Bishop's Stortford, Herts CM23 3DH, UK.

Oncor Inc., 209 Perry Parkway, Gaithersburg, MD 20877, USA.

Oxoid, *see* **Unipath Ltd**.

Pharmacia LKB Biotechnology AB, Björkgatan 30, Uppsala 75182, Sweden; Pharmacia Biosystems Ltd, Davy Ave, Knowhill, Milton Keynes MK9 3HP, UK.

Polaroid Corporation, Cambridge, MA 01239, USA.

Promega Corporation, 2800 Woods Hollow Road, Madison, WI 53711–5399, USA.

Serva Biochemicals Inc., 200 Shames Drive, Westbury, NY 11590, USA.

Sigma Chemical Company, PO Box 14509, St. Louis MO 63178, USA; Sigma Chemical Company Ltd, Fancy Road, Poole, Dorset BH17 7NH, UK.

Stratagene, 11011 North Torrey Pines Road, La Jolla, CA 92037, USA.

Unipath Ltd, Wade Road, Basingstoke, Hampshire RG24 8PW, UK.

Vector Laboratories Inc., 30 Ingold Road, Burlingame, CA 94010, USA.

Whatman Laboratory Products Inc., 9 Bridewell Place, Clifton, NJ 07014, USA; Whatman Ltd, Springfield Mill, Maidstone, Kent ME14 2LE, UK.

Index